Discrete and Computational
GEOMETRY

Discrete and Computational
GEOMETRY

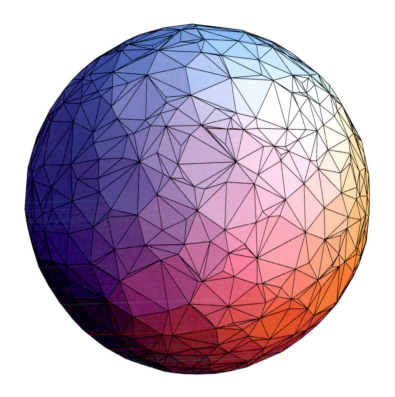

SATYAN L. DEVADOSS

and

JOSEPH O'ROURKE

PRINCETON UNIVERSITY PRESS
PRINCETON AND OXFORD

Published by Princeton University Press,
41 William Street, Princeton, New Jersey 08540
In the United Kingdom: Princeton University Press,
6 Oxford Street, Woodstock, Oxfordshire OX20 1TW

Library of Congress Cataloging-in-Publication Data

Devadoss, Satyan L., 1973–
Discrete and computational geometry / Satyan L. Devadoss and Joseph O'Rourke.
   p. cm.
Includes index.
ISBN 978-0-691-14553-2 (hardcover : alk. paper)
1. Geometry–Data processing.   I. O'Rourke, Joseph.   II. Title.
QA448.D38D48 2011
516.00285–dc22   2010044434

British Library Cataloging-in-Publication Data is available

This book has been composed in Sabon

Princeton University Press books are printed on acid-free paper, and meet the guidelines for permanence and durability of the Committee on Production Guidelines for Book Longevity of the Council on Library Resources

Typeset by S R Nova Pvt Ltd, Bangalore, India

press.princeton.edu

Printed in Thailand

10 9 8 7 6 5 4 3 2 1

Although geometry is as old as mathematics itself, discrete geometry only fully emerged in the twentieth century, and computational geometry was only christened in the late 1970s. The terms "discrete" and "computational" fit well together, as the geometry must be discretized in preparation for computations. "Discrete" here means concentration on finite sets of points, lines, triangles, and other geometric objects, and is used to contrast with "continuous" geometry, for example, smooth surfaces. Although the two endeavors were growing naturally on their own, it has been the interaction between discrete and computational geometry that has generated the most excitement, with each advance in one field spurring an advance in the other. The interaction also draws upon two traditions: theoretical pursuits in pure mathematics and applications-driven directions often arising in computer science. The confluence has made the topic an ideal bridge between mathematics and computer science. It is precisely to bridge that gap that we have written this book.

In line with this goal, our presentation is sprinkled with both algorithms and theorems, with sometimes the theorem serving as the main thrust (e.g., the Gauss-Bonnet theorem), and sometimes an algorithm the primary goal of a section and theorems playing a supporting role (e.g., the flip graph computation of the Delaunay triangulation). As our emphasis is on the geometry of the subject, the algorithms presented in this book are strongly rooted in geometric intuition and insight. We describe the algorithms independent of any particular programming language, and in fact we do not even employ pseudocode, trusting that our boxed descriptions can be read as code by those steeped in the computer science idiom. Thus, no programming experience is needed to read this book. Algorithm complexities are discussed using the big-Oh notation without an assumption of prior exposure to this style of thinking, which is (lightly) covered in the Appendix.

We include many proofs that we feel would interest a mathematically inclined student, presenting them in what we hope is an accessible style. At many junctures we connect to more advanced concerns, not fearing, for example, to jump to higher dimensions to make a relevant remark. At the same time, we connect each topic to applications that were often the initial motivation for studying the topic. Although we include careful proofs of theorems, we also try to develop intuition through visualization. Geometry demands figures!

Some exposure to proofs is needed to gain that mystical "mathematical maturity." We invoke calculus only in a few sections. A course in discrete mathematics is the more relevant prerequisite, but any course that presents formal proofs of theorems would suffice, such as linear algebra or automata theory. The material should be completely accessible to any mathematics or computer science major in the second or third year of college. In order to reach interesting advanced topics without the careful preparation they often demand, we sometimes offer a *proof sketch* (always marked as such), instead of a long, detailed formal proof of a result. Here we try to convince the reader that a formal proof is likely to be possible by sketching in the outlines without the details. Whether the reader can imagine those details from the outline is a measure of mathematical experience. A parallel skill of "computational maturity" is needed to imagine how to implement our algorithm descriptions.

The book is studded with Exercises, which we have chosen to place wherever they are relevant, rather than gather them at the end of each chapter. Some merely test a grasp of the foregoing material, most require more substantive thought (suitable for homework assignments), and starred exercises ★ are difficult, often connecting to a published paper. A solutions manual is available to instructors from the publisher.

Rather than include a scholarly bibliography, we have opted instead for "Suggested Readings" at the end of each chapter, providing pointers for further investigation. Between these pointers and websites such as Wikipedia, the reader should have no difficulty exploring the vast area beyond our coverage. And what lies beyond is indeed vast. The *Handbook of Discrete and Computational Geometry* runs to 1,500 pages and even so is highly compressed. Our coverage represents a sparse sampling of the field. We have chosen to cover polygons, convex hulls, triangulations, and Voronoi diagrams, which we believe constitute the core of discrete and computational geometry. Beyond this core, there is considerable choice, and we have selected several topics on curves and polyhedra, concluding with configuration spaces. The selection is skewed to the research interests of the authors, with perhaps more coverage of associahedra (first author) and unfolding (second author) than might be chosen by a committee of our peers. At the least, this ensures that we touch on the frontiers of current research.

Because of the relative youth of the field, there are many accessible unsolved problems, which we highlight throughout. Although some have resisted the assaults of many talented researchers and may be awaiting a theoretical breakthrough, others may be accessible with current techniques and only await significant attention by an enterprising reader.

The field has expanded greatly since its origins, and the new connections to areas of mathematics (such as algebraic topology) and new application areas (such as data mining) seems only to be accelerating. We hope this book can serve to open the door on this rich and fascinating subject.

———————————————

Acknowledgments. Vickie Kearn was the ideal editor for us: firm but kind, and unfailingly enthusiastic. A special thanks goes to wise Jeff Erickson, who read the entire manuscript in draft, corrected many errors, suggested many exercises, and in general *educated* us in our own specialties to a degree we did not think possible. We are humbled and grateful.

Satyan Devadoss: To all my students at Williams College who have learned this beautiful subject alongside me, while enduring my brutal exams, I am truly grateful. I especially thank Katie Baldiga, Jeff Danciger, Thomas Kindred, Rohan Mehra, Nick Perry, and Don Sheehy for being on the front line with me, with special thanks to Tomio Ueda for literally laying the foundation to this book.

I am indebted to my colleagues and mentors Colin Adams, Mike Davis, Tamal Dey, Peter March, Jack Morava, Frank Morgan, Alan Saalfeld, and Jim Stasheff, all of whom have given generously of their time and wisdom over the years. Thanks also go to the Ohio State University, the Mathematical Sciences Research Institute, and the University of California, Berkeley, for their hospitality during my sabbatical visits where parts of this book were written. The NSF and DARPA were also instrumental by their support with grant DMS-0310354.

Joseph O'Rourke: I thank my students Nadia Benbernou, Julie DiBiase, Melody Donoso, Biliana Kaneva, Anna Lysyanskaya, Stacia Wyman, and Dianna Xu, all of whose work found its way into the book in one form or another. I thank my colleagues and coauthors Lauren Cowles, Erik Demaine, Jin-ichi Ito, Joseph Mitchell, Don Shimamoto, and Costin Vîlcu, who each taught me so much through our collaborations. The early stages of my work on this book were funded by a NSF Distinguished Teaching Scholars award DUE-0123154.

Satyan L. Devadoss
Williams College

Joseph O'Rourke
Smith College

Discrete and Computational
GEOMETRY

POLYGONS 1

Polygons are to planar geometry as integers are to numerical mathematics: a discrete subset of the full universe of possibilities that lends itself to efficient computations. And triangulations are the prime factorizations of polygons, alas without the benefit of the "Fundamental Theorem of Arithmetic" guaranteeing unique factorization. This chapter introduces triangulations (Section 1.1) and their combinatorics (Section 1.2), and then applies these concepts to the alluring art gallery theorem (Section 1.3), a topic at the roots of computational geometry which remains an active area of research today. Here we encounter a surprising difference between 2D triangulations and 3D tetrahedralizations.

Triangulations are highly constrained decompositions of polygons. Dissections are less constrained partitions, and engender the fascinating question of which pairs of polygons can be dissected and reassembled into each other. This so-called "scissors congruence" (Section 1.4) again highlights the fundamental difference between 2D and 3D (Section 1.5), a theme throughout the book.

1.1 DIAGONALS AND TRIANGULATIONS

Computational geometry is fundamentally *discrete* as opposed to continuous. Computation with curves and smooth surfaces are generally considered part of another field, often called "geometric modeling." The emphasis on computation leads to a focus on representations of geometric objects that are simple and easily manipulated. Fundamental building blocks are the *point* and the line *segment*, the portion of a line between two points. From these are built more complex structures. Among the most important of these structures are 2D polygons and their 3D generalization, polyhedra.

A *polygon*[1] P is the closed region of the plane bounded by a finite collection of line segments forming a closed curve that does not intersect itself. The line segments are called *edges* and the points where adjacent edges meet are called *vertices*. In general, we insist that vertices be true corners at which there is a bend between the adjacent edges, but in some

---

[1] Often the term *simple polygon* is used, to indicate that it is "simply connected," a concept we explore in Chapter 5.

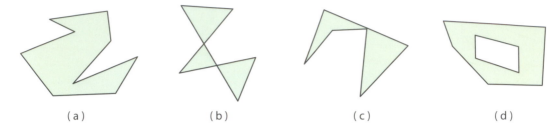

(a) (b) (c) (d)

Figure 1.1. (a) A polygon. (b)–(d) Objects that are not polygons.

circumstances (such as in Chapter 2) it will be useful to recognize "flat vertices." The set of vertices and edges of P is called the *boundary* of the polygon, denoted as ∂P. Figure 1.1(a) shows a polygon with nine edges joined at nine vertices. Diagrams (b)–(d) show objects that fail to be polygons.

The fundamental "Jordan curve theorem," formulated and proved by Camille Jordan in 1882, is notorious for being both obvious and difficult to prove in its full generality. For polygons, however, the proof is easier, and we sketch the main idea.

Theorem 1.1 (Polygonal Jordan Curve). *The boundary ∂P of a polygon P partitions the plane into two parts. In particular, the two components of $\mathbb{R}^2 \setminus \partial P$ are the bounded interior and the unbounded exterior.*[2]

Sketch of Proof. Let P be a polygon in the plane. We first choose a fixed direction in the plane that is not parallel to any edge of P. This is always possible because P has a finite number of edges. Then any point x in the plane not on ∂P falls into one of two sets:

1. The ray through x in the fixed direction crosses ∂P an even number of times: x is exterior. Here a ray through a vertex is not counted as crossing ∂P.
2. The ray through x in the fixed direction crosses ∂P an odd number of times: x is interior.

Notice that all points on a line segment that do not intersect ∂P must lie in the same set. Thus the even sets and the odd sets are connected. And moreover, if there is a path between points in different sets, then this path must intersect ∂P. □

This proof sketch is the basis for an algorithm for deciding whether a given point is inside a polygon, a low-level task that is encountered every time a user clicks inside some region in a computer game, and in many other applications.

---

[2] The symbol '$\setminus$' indicates set subtraction: $A \setminus B$ is the set of points in A but not in B.

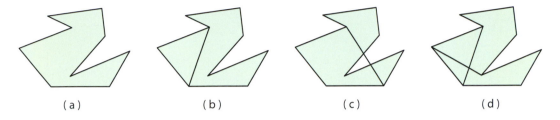

Figure 1.2. (a) A polygon with (b) a diagonal; (c) a line segment; (d) crossing diagonals.

Exercise 1.2. *Flesh out the proof of Theorem 1.1 by supplying arguments to (a) justify the claim that if there is a path between the even- and odd-crossings sets, the path must cross ∂P; and (b) establish that for two points in the same set, there is a path connecting them that does not cross ∂P.*

Algorithms often need to break polygons into pieces for processing. A natural decomposition of a polygon P into simpler pieces is achieved by drawing diagonals. A *diagonal* of a polygon is a line segment connecting two vertices of P and lying in the interior of P, not touching ∂P except at its endpoints. Two diagonals are *noncrossing* if they share no interior points. Figure 1.2 shows (a) a polygon, (b) a diagonal, (c) a line segment that is not a diagonal, and (d) two crossing diagonals.

Definition. A *triangulation* of a polygon P is a decomposition of P into triangles by a maximal set of noncrossing diagonals.

Here *maximal* means that no further diagonal may be added to the set without crossing (sharing an interior point with) one already in the set. Figure 1.3 shows a polygon with three different triangulations. Triangulations lead to several natural questions. How many different triangulations does a given polygon have? How many triangles are in each triangulation of a given polygon? Is it even true that every polygon always *has* a triangulation? Must every polygon have at least one diagonal? We start with the last question.

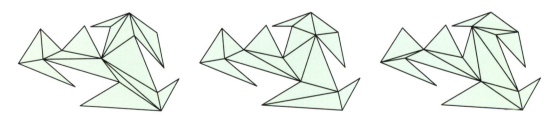

Figure 1.3. A polygon and three possible triangulations.

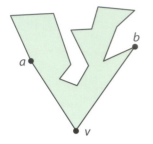 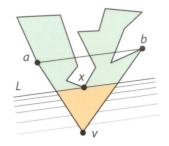 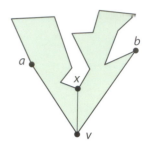

Figure 1.4. Finding a diagonal of a polygon through sweeping.

Lemma 1.3. *Every polygon with more than three vertices has a diagonal.*

Proof. Let v be the lowest vertex of P; if there are several, let v be the rightmost. Let a and b be the two neighboring vertices to v. If the segment ab lies in P and does not otherwise touch ∂P, it is a diagonal. Otherwise, since P has more than three vertices, the closed triangle formed by a, b, and v contains at least one vertex of P. Let L be a line parallel to segment ab passing through v. Sweep this line from v parallel to itself upward toward ab; see Figure 1.4. Let x be the first vertex in the closed triangle abv, different from a, b, or v, that L meets along this sweep. The (shaded) triangular region of the polygon below line L and above v is empty of vertices of P. Because vx cannot intersect ∂P except at v and x, we see that vx is a diagonal. $\square$

Since we can decompose any polygon (with more than three vertices) into two smaller polygons using a diagonal, induction leads to the existence of a triangulation.

Theorem 1.4. *Every polygon has a triangulation.*

Proof. We prove this by induction on the number of vertices n of the polygon P. If $n = 3$, then P is a triangle and we are finished. Let $n > 3$ and assume the theorem is true for all polygons with fewer than n vertices. Using Lemma 1.3, find a diagonal cutting P into polygons P_1 and P_2. Because both P_1 and P_2 have fewer vertices than n, P_1 and P_2 can be triangulated by the induction hypothesis. By the Jordan curve theorem (Theorem 1.1), the interior of P_1 is in the exterior of P_2, and so no triangles of P_1 will overlap with those of P_2. A similar statement holds for the triangles of P_2. Thus P has a triangulation as well. $\square$

Exercise 1.5. *Prove that every polygonal region with polygonal holes, such as Figure 1.1(d), admits a triangulation of its interior.*

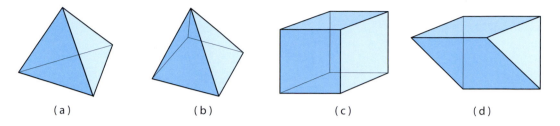

Figure 1.5. Polyhedra: (a) tetrahedron, (b) pyramid with square base, (c) cube, and (d) triangular prism.

That every polygon has a triangulation is a fundamental property that pervades discrete geometry and will be used over and over again in this book. It is remarkable that this notion does not generalize smoothly to three dimensions. A *polyhedron* is the 3D version of a polygon, a 3D solid bounded by finitely many polygons. Chapter 6 will define polyhedra more precisely and explore them more thoroughly. Here we rely on intuition. Figure 1.5 gives examples of polyhedra.

Just as the simplest polygon is the triangle, the simplest polyhedron is the *tetrahedron*: a pyramid with a triangular base. We can generalize the 2D notion of polygon triangulation to 3D: a *tetrahedralization* of a polyhedron is a partition of its interior into tetrahedra whose edges are diagonals of the polyhedron. Figure 1.6 shows examples of tetrahedralizations of the polyhedra just illustrated.

Exercise 1.6. *Find a tetrahedralization of the cube into five tetrahedra.*

We proved in Theorem 1.4 that all polygons can be triangulated. Does the analogous claim hold for polyhedra: can all polyhedra be

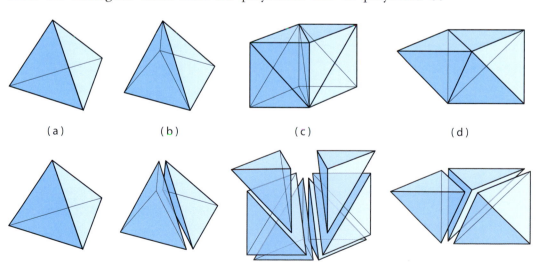

Figure 1.6. Tetrahedralizations of the polyhedra from Figure 1.5.

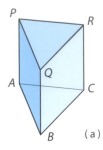 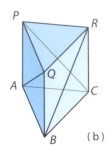 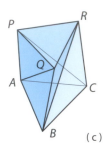 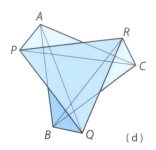

Figure 1.7. Construction of the Schönhardt polyhedron from a triangular prism, where (d) is the overhead view.

tetrahedralized? In 1911, Nels Lennes proved the surprising theorem that this is not so. We construct an example of a polyhedron, based on the 1928 model by Erich Schönhardt, which cannot be tetrahedralized. Let A, B, C be vertices of an equilateral triangle (labeled counterclockwise) in the xy-plane. Translating this triangle vertically along the z-axis reaching $z = 1$ traces out a triangular prism, as shown in Figure 1.7(a). Part (b) shows the prism with the faces partitioned by the diagonal edges AQ, BR, and CP. Now twist the top PQR triangle $\pi/6$ degrees in the $(z = 1)$-plane, rotating and stretching the diagonal edges. The result is the Schönhardt polyhedron, shown in (c) and in an overhead view in (d) of the figure. Schönhardt proved that this is the smallest example of an untetrahedralizable polyhedron.

Exercise 1.7. *Prove that the Schönhardt polyhedron cannot be tetrahedralized.*

UNSOLVED PROBLEM 1 Tetrahedralizable Polyhedra

Find characteristics that determine whether or not a polyhedron is tetrahedralizable. Even identifying a large natural class of tetrahedralizable polyhedra would be interesting.

This is indeed a difficult problem. It was proved by Jim Ruppert and Raimund Seidel in 1992 that it is NP-complete to determine whether a polyhedron is tetrahedralizable. *NP-complete* is a technical term from complexity theory that means, roughly, an intractable algorithmic problem. (See the Appendix for a more thorough explanation.) It suggests in this case that there is almost certainly no succinct characterization of tetrahedralizability.

1.2 BASIC COMBINATORICS

We know that every polygon has at least one triangulation. Next we show that the number of triangles in any triangulation of a fixed polygon is the same. The proof is essentially the same as that of Theorem 1.4, with more quantitative detail.

Theorem 1.8. *Every triangulation of a polygon P with n vertices has $n - 2$ triangles and $n - 3$ diagonals.*

Proof. We prove this by induction on n. When $n = 3$, the statement is trivially true. Let $n > 3$ and assume the statement is true for all polygons with fewer than n vertices. Choose a diagonal d joining vertices a and b, cutting P into polygons P_1 and P_2 having n_1 and n_2 vertices, respectively. Because a and b appear in both P_1 and P_2, we know $n_1 + n_2 = n + 2$. The induction hypothesis implies that there are $n_1 - 2$ and $n_2 - 2$ triangles in P_1 and P_2, respectively. Hence P has

$$(n_1 - 2) + (n_2 - 2) = (n_1 + n_2) - 4 = (n + 2) - 4 = n - 2$$

triangles. Similarly, P has $(n_1 - 3) + (n_2 - 3) + 1 = n - 3$ diagonals, with the $+1$ term counting d.   □

Many proofs and algorithms that involve triangulations need a special triangle in the triangulation to initiate induction or start recursion. "Ears" often serve as special triangles. Three consecutive vertices a, b, c form an *ear* of a polygon if ac is a diagonal of the polygon. The vertex b is called the ear *tip*.

Corollary 1.9. *Every polygon with more than three vertices has at least two ears.*

Proof. Consider any triangulation of a polygon P with $n > 3$ vertices, which by Theorem 1.8 partitions P into $n - 2$ triangles. Each triangle covers at most two edges of ∂P. Because there are n edges on the boundary of P but only $n - 2$ triangles, by the pigeonhole principle at least two triangles must contain two edges of P. These are the ears.   □

Exercise 1.10. *Prove Corollary 1.9 using induction.*

Exercise 1.11. *Show that the sum of the interior angles of any polygon with n vertices is $\pi(n - 2)$.*

Exercise 1.12. *Using the previous exercise, show that the total turn angle around the boundary of a polygon is 2π. Here the turn angle at a vertex v is π minus the internal angle at v.*

Exercise 1.13. *Three consecutive vertices a, b, c form a mouth of a polygon if ac is an external diagonal of the polygon, a segment wholly outside. Formulate and prove a theorem about the existence of mouths.*

Exercise 1.14. *Let a polygon P with h holes have n total vertices (including hole vertices). Find a formula for the number of triangles in any triangulation of P.*

★ **Exercise 1.15.** *Let P be a polygon with vertices (x_i, y_i) in the plane. Prove that the area of P is*

$$\frac{1}{2} \left| \sum (x_i y_{i-1} - x_{i-1} y_i) \right|.$$

Although the number of triangles in any triangulation of a polygon is the same, it is natural to explore the number of different triangulations of a given polygon. For instance, Figure 1.3 shows a polygon with three different triangulations.

Exercise 1.16. *For each polygon in Figure 1.8, find the number of distinct triangulations.*

Exercise 1.17. *For each $n > 3$, find a polygon with n vertices that has a unique triangulation.*

The number of triangulations of a fixed polygon P has much to do with the "shape" of the polygon. One crucial measure of shape is the internal angles at the vertices. A vertex of P is called *reflex* if its angle is greater than π, and *convex* if its angle is less than or equal to π. Sometimes it is useful to distinguish a *flat* vertex, whose angle is exactly π, from a *strictly convex* vertex, whose angle is strictly less than π. A polygon P is a *convex* polygon if all vertices of P are convex. In general we exclude flat vertices, so unless otherwise indicated, the vertices of a convex polygon

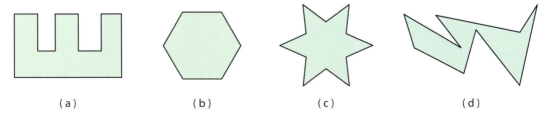

(a)   (b)   (c)   (d)

Figure 1.8. Find the number of distinct triangulations for each of the polygons given.

are strictly convex. With this understanding, a convex polygon has the following special property.

Lemma 1.18. *A diagonal exists between any two nonadjacent vertices of a polygon P if and only if P is a convex polygon.*

Proof. The proof is in two parts, both established by contradiction. First assume P is not convex. We need to find two vertices of P that do not form a diagonal. Because P is not convex, there exists a sequence of three vertices a, b, c, with b reflex. Then the segment ac lies (at least partially) exterior to P and so is not a diagonal.

Now assume P is convex but there are a pair of vertices a and b in P that do not form a diagonal. We identify a reflex vertex of P to establish the contradiction. Let σ be the shortest path connecting a to b entirely within P. It cannot be that σ is a straight segment contained inside P, for then ab is a diagonal. Instead, σ must be a chain of line segments. Each corner of this polygonal chain turns at a reflex vertex — if it turned at a convex vertex or at a point interior to P, it would not be the shortest. □

For a convex polygon P, where every pair of nonadjacent vertices determines a diagonal, it is possible to count the number of triangulations of P based solely on the number of vertices. The result is the *Catalan number*, named after the nineteenth-century Belgian mathematician Eugène Catalan.

Theorem 1.19. *The number of triangulations of a convex polygon with $n + 2$ vertices is the Catalan number*

$$C_n = \frac{1}{n+1}\binom{2n}{n}. \tag{1.1}$$

Proof. Let P_{n+2} be a convex polygon with vertices labeled from 1 to $n + 2$ counterclockwise. Let $\mathcal{T}_{n+2}$ be the set of triangulations of P_{n+2}, where $\mathcal{T}_{n+2}$ has t_{n+2} elements. We wish to show that t_{n+2} is the Catalan number C_n.

Let ϕ be the map from $\mathcal{T}_{n+2}$ to $\mathcal{T}_{n+1}$ given by contracting the edge $\{1, n + 2\}$ of P_{n+2}. To *contract* an edge ab is to shrink it to a point c so that c becomes incident to all the edges and diagonals that were incident to either a or b. Let T be an element of $\mathcal{T}_{n+1}$. What is important to note is the number of triangulations of $\mathcal{T}_{n+2}$ that map to T (i.e., the number of elements of $\phi^{-1}(T)$) equals the degree of vertex 1 in T. Figure 1.9 gives an example where (a) five triangulations of the octagon all map to (b) the same triangulation of the heptagon, where the vertex labeled 1 has degree five. This is evident since each edge incident to 1 can *open up* into a triangle in $\phi^{-1}(T)$, shown by the shaded triangles

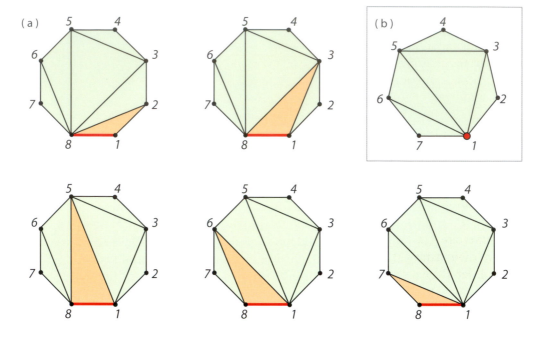

Figure 1.9. The five polygons in (a) all map to the same polygon in (b) under contraction of edge $\{1, 8\}$.

in (a). So we see that

$$t_{n+2} = \sum_{T \in \mathcal{T}_{n+1}} \text{degree of vertex 1 of } T.$$

Because this polygon is convex, this is true for all vertices of T. Therefore we can sum over all vertices of T, obtaining

$$(n+1) \cdot t_{n+2} = \sum_{i=1}^{n+1} \sum_{T \in \mathcal{T}_{n+1}} \text{degree of vertex } i \text{ of } T$$

$$= \sum_{T \in \mathcal{T}_{n+1}} \sum_{i=1}^{n+1} \text{degree of vertex } i \text{ of } T$$

$$= 2(2n-1) \cdot t_{n+1}.$$

The last equation follows because the sum of the degrees of all vertices of T double-counts the number of edges of T and the number of diagonals of T. Because T is in $\mathcal{T}_{n+1}$, it has $n+1$ edges, and by Theorem 1.8, it has $n-2$ diagonals. Solving for t_{n+2}, we get

$$t_{n+2} = \frac{2(2n-1)}{n+1} \cdot t_{n+1} = 2^n \cdot \frac{2n-1}{n+1} \cdot \frac{2n-3}{n} \cdots \frac{3}{3} \cdot \frac{1}{2}$$

$$= \frac{(2n)!}{(n+1)! \, n!} = \frac{1}{n+1} \binom{2n}{n}.$$

This is the Catalan number C_n, completing the proof.   □

For the octagon in Figure 1.9, the formula shows there are $C_6 = 132$ distinct triangulations. Is it possible to find a closed formula for the number of triangulations for nonconvex polygons P with n vertices? The answer, unfortunately, is NO, because small changes in the position of vertices can lead to vastly different triangulations of the polygon. What we do know is that convex polygons achieve the maximum number of triangulations.

Theorem 1.20. *Let P be a polygon with $n + 2$ vertices. The number of triangulations of P is between 1 and C_n.*

Proof. Exercise 1.17 shows there are polygons with exactly one triangulation, demonstrating that the lower bound is realizable. For the upper bound, let P be any polygon with n labeled, ordered vertices, and let Q be a convex polygon also with n vertices, labeled similarly. Each diagonal of P corresponds to a diagonal of Q, and if two diagonals of P do not cross, neither do they cross in Q. So every triangulation of P (having $n - 1$ diagonals by Theorem 1.8) determines a triangulation of Q (again with $n - 1$ diagonals). Therefore P can have no more triangulations than Q, which by Theorem 1.19 is C_n.   □

Thus we see that convex polygons yield the most triangulations. Because convex polygons have no reflex vertices (by definition), there might possibly be a relationship between the number of triangulations and the number of reflex vertices of a polygon. Sadly, this is not the case. Let P be a polygon with five vertices. By Theorem 1.19, if P has no reflex vertices, it must have 5 triangulations. Figure 1.10(a) shows P with one reflex vertex and only one triangulation, whereas parts (b) and (c) show P with two reflex vertices and two triangulations. So the number of triangulations does not necessarily decrease with the number of reflex vertices. In fact, the number of triangulations does not depend on the number of reflex vertices at all. Figure 1.10(d) shows a polygon with a unique triangulation with three reflex vertices. This example can be generalized to polygons with unique triangulations that contain arbitrarily many reflex vertices.

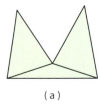
(a)   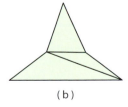 (b)   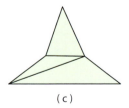 (c)   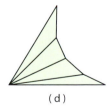 (d)

Figure 1.10. Triangulations of special polygons.

Exercise 1.21. *For each n > 3, find a polygon with n vertices with exactly two triangulations.*

Exercise 1.22. *For any n ≥ 3, show there is no polygon with n + 2 vertices with exactly $C_n - 1$ triangulations.*

UNSOLVED PROBLEM 2   Counting Triangulations

Identify features of polygons P that lead to a closed formula for the number of triangulations of P in terms of those features.

We learned earlier that properties can be lost in the move from 2D polygons to 3D polyhedra. For example, all polygons can be triangulated but not all polyhedra can be tetrahedralized. Moreover, by Theorem 1.8 above, we know that *every* polygon with n vertices must have the same number of triangles in *any* of its triangulation. For polyhedra, this is far from true. In fact, two different tetrahedralizations of the *same* polyhedron can result in a different number of tetrahedra! Consider Figure 1.11, which shows a polyhedron partitioned into two tetrahedra (a) and also into three (b).

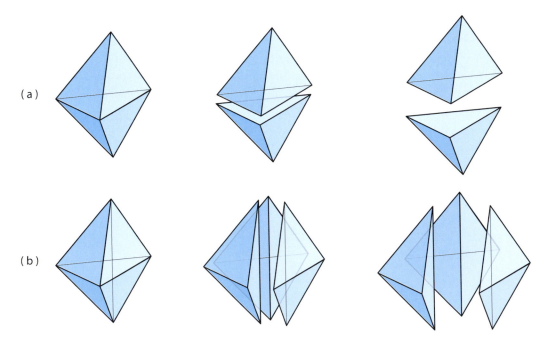

(a)

(b)

Figure 1.11. A polyhedron partitioned into (a) two and (b) three tetrahedra.

Even for a polyhedron as simple as the cube, the number of tetrahedra is not the same for all tetrahedralizations. It turns out that up to rotation and reflection, there are six different tetrahedralizations of the cube, one of which was shown earlier in Figure 1.6(c). Five of the six partition the cube into six tetrahedra, but one cuts it into only five tetrahedra.

Exercise 1.23. *Is it possible to partition a cube into six congruent tetrahedra? Defend your answer.*

Exercise 1.24. *Find the six different tetrahedralizations of the cube up to rotation and reflection.*

★ **Exercise 1.25.** *Classify the set of triangulations on the boundary of the cube that "induce" tetrahedralizations of the cube, where each such tetrahedralization matches the triangulation on the cube surface.*

As is common in geometry, concepts that apply to 2D and to 3D generalize to arbitrary dimensions. The *n*-dimensional generalization of the triangle/tetrahedron is the *n-simplex* of $n + 1$ vertices. Counting *n*-dimensional "triangulations" is largely unsolved:

UNSOLVED PROBLEM 3   Simplices and Cubes

Find the smallest triangulation of the *n*-dimensional cube into *n*-simplices. It is known, for example, that the 4D cube (the *hypercube*) may be partitioned into 16 4-simplices, and this is minimal. But the minimum number is unknown except for the few small values of *n* that have yielded to exhaustive computer searches.

Exercise 1.26. *Show that the n-dimensional cube can be triangulated into exactly n! simplices.*

1.3 THE ART GALLERY THEOREM

A beautiful problem posed by Victor Klee in 1973 engages several of the concepts we have discussed: Imagine an art gallery whose floor plan is modeled by a polygon. A guard of the gallery corresponds to a point on our polygonal floor plan. Guards can see in every direction, with a full 360° range of visibility. Klee asked to find the fewest number of (stationary) guards needed to protect the gallery. Before tackling this problem, we need to define what it means to "see something" mathematically.

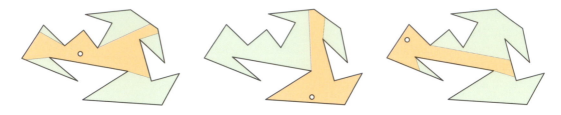

Figure 1.12. Examples of the range of visibility available to certain placement of guards.

A point x in polygon P is *visible* to point y in P if the line segment xy lies in P. This definition allows the line of sight to have a grazing contact with the boundary ∂P (unlike the definition for diagonal). A set of guards *covers* a polygon if every point in the polygon is visible to some guard. Figure 1.12 gives three examples of the range of visibility available to single guards in polygons.

A natural question is to ask for the *minimum* number of guards needed to cover polygons. Of course, this minimum number depends on the "complexity" of the polygon in some way. We choose to measure complexity in terms of the number of vertices of the polygon. But two polygons with n vertices can require different numbers of guards to cover them. Thus we look for a bound that is good for *any* polygon with n vertices.[3]

Exercise 1.27. *For each polygon in Figure 1.8, find the minimum number of guards needed to cover it.*

Exercise 1.28. *Suppose that guards themselves block visibility so that a line of sight from one guard cannot pass through the position of another. Are there are polygons for which the minimum of our more powerful guards needed is strictly less than the minimum needed for these weaker guards?*

Let's start by looking at some examples for small values of n. Figure 1.13 shows examples of covering guard placements for polygons with a small number of vertices. Clearly, any triangle only needs one guard to cover it. A little experimentation shows that the first time two guards are needed is for certain kinds of hexagons.

Exercise 1.29. *Prove that any quadrilateral needs only one guard to cover it. Then prove that any pentagon needs only one guard to cover it.*

---

[3] To find the minimum number of guards for a particular polygon turns out to be, in general, an intractable algorithmic task. This is an instance of another NP-complete problem; see the Appendix.

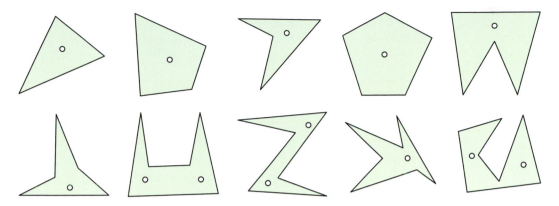

Figure 1.13. Examples of guard placements for different polygons.

Exercise 1.30. *Modify Lemma 1.18 to show that one guard placed anywhere in a convex polygon can cover it.*

By the previous exercise, convex polygons need only one guard for coverage. The converse of this statement is not true, however. There are polygons that need only one guard but which are not convex. These polygons are called *star* polygons. Figure 1.8(c) is an example of a star polygon.

While correct placement avoids the need for a second guard for quadrilaterals and pentagons, one can begin to see how reflex vertices will cause problems in polygons with large numbers of vertices. Because there can exist only so many reflex angles in a polygon, we can construct a useful example, based on prongs. Figure 1.14 illustrates the comb-shaped design made of 5 prongs and 15 vertices. We can see that a comb of n prongs has $3n$ vertices, and since each prong needs its own guard, then at least $\lfloor n/3 \rfloor$ guards are needed. Here the symbols $\lfloor\ \rfloor$ indicate the *floor* function: the largest integer less than or equal to the enclosed argument.[4] Thus we have a lower bound on Klee's problem: $\lfloor n/3 \rfloor$ guards are sometimes necessary.

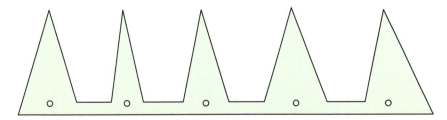

Figure 1.14. A comb-shaped example.

---

[4] Later we will use its cousin, the *ceiling* function $\lceil\ \rceil$, the smallest integer greater than or equal to the argument.

Exercise 1.31. *Construct a polygon P and a placement of guards such that the guards see every point of ∂P but P is not covered.*

UNSOLVED PROBLEM 4   Visibility Graphs

The *visibility graph* of a polygon P is the graph with a node for each vertex of P and an arc connecting two nodes when the corresponding vertices of P can see one another. Find necessary and sufficient conditions that determine when a graph is the visibility graph of some polygon.

Now that we have a lower bound of $\lfloor n/3 \rfloor$, the next question is whether this number always suffices, that is, is it also an upper bound for all polygons? Other than proceeding case by case, how can we attack the problem from a general framework? The answer lies in triangulating the polygon. Theorem 1.4 implies that every polygon with n vertices can be covered with $n - 2$ guards by placing a guard in each triangle, providing a crude upper bound. But we have been able to do better than this already for quadrilaterals and pentagons. By placing guards not *in* each triangle but on the *vertices*, we can possibly cover more triangles by fewer guards. In 1975, Vašek Chvátal found a proof for the minimum number of guards needed to cover any polygon with n vertices. His proof is based on induction, with some delicate case analysis. A few years later, Steve Fisk found another, beautiful inductive proof, which follows below.

Theorem 1.32 (Art Gallery). *To cover a polygon with n vertices, $\lfloor n/3 \rfloor$ guards are needed for some polygons, and sufficient for all of them.*

Proof. We already saw in Figure 1.14 that $\lfloor n/3 \rfloor$ guards can be necessary. We now need to show this number also suffices.

Consider a triangulation of a polygon P. We use induction to prove that each vertex of P can be assigned one of three colors (i.e., the triangulation can be 3-*colored*), so that any pair of vertices connected by an edge of P or a diagonal of the triangulation must have different colors. This is certainly true for a triangle. For $n > 3$, Corollary 1.9 guarantees that P has an ear abc, with vertex b as the ear tip. Removing the ear produces a polygon P' with $n - 1$ vertices, where b has been removed. By the induction hypothesis, the vertices of P' can be 3-colored. Replacing the ear, coloring b with the color not used by a or c, provides a coloring for P.

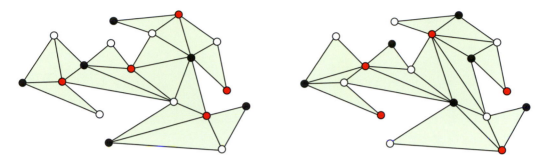

Figure 1.15. Triangulations and colorings of vertices of a polygon with $n = 18$ vertices. In both figures, red is the least frequently used color, occurring five times.

Since there are n vertices, by the pigeonhole principle, the least frequently used color appears on at most $\lfloor n/3 \rfloor$ vertices. Place guards at these vertices. Figure 1.15 shows two examples of triangulations of a polygon along with colorings of the vertices as described. Because every triangle has one corner a vertex of this color, and this guard covers the triangle, the museum is completely covered. □

Exercise 1.33. *For each polygon in Figure 1.16, find a minimal set of guards that cover it.*

Exercise 1.34. *Construct a polygon with $n = 3k$ vertices such that placing a guard at every third vertex fails to protect the gallery.*

The classical art gallery problem as presented has been generalized in several directions. Some of these generalizations have elegant solutions, some have difficult solutions, and several remain unsolved problems. For instance, the shape of the polygons can be restricted (to polygons with right-angled corners) or enlarged (to include polygons with holes), or the mobility of the guards can be altered (permitting guards to walk along edges, or along diagonals).

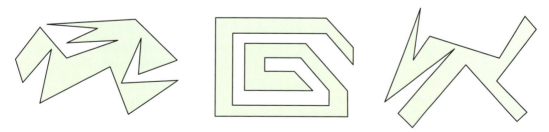

Figure 1.16. Find a set of minimal guards that cover the polygons.

Exercise 1.35. *Why is it not possible to easily extend Fisk's proof above to the case of polygons with holes?*

Exercise 1.36. *Using Exercise 1.14, derive an upper bound on the number of guards needed to cover a polygon with h holes and n total vertices. (Obtaining a tight upper bound is extremely difficult, and only recently settled.)*

When all edges of the polygon meet at right angles (an *orthogonal* polygon), fewer guards are needed, as established by Jeff Kahn, Maria Klawe, and Daniel Kleitman in 1980. In contrast, covering the exterior rather than the interior of a polygon requires (in general) more guards, established by Joseph O'Rourke and Derick Wood in 1983.

Theorem 1.37 (Orthogonal Gallery). *To cover polygons with n vertices with only right-angled corners, $\lfloor n/4 \rfloor$ guards are needed for some polygons, and sufficient for all of them.*

Theorem 1.38 (Fortress). *To cover the* exterior *of polygons with n vertices, $\lceil n/2 \rceil$ guards are needed for some polygons, and sufficient for all of them.*

Exercise 1.39. *Prove the Fortress theorem.*

Exercise 1.40. *For any $n > 3$, construct a polygon P with n vertices such that $\lceil n/3 \rceil$ guards, placed anywhere on the plane, are sometimes necessary to cover the exterior of P.*

UNSOLVED PROBLEM 5   Edge Guards

An *edge guard* along edge e of polygon P sees a point y in P if there exists x in e such that x is visible to y. Find the number of edge guards that suffice to cover a polygon with n vertices. Equivalently, how many edges, lit as fluorescent bulbs, suffice to illuminate the polygon? Godfried Toussaint conjectured that $\lfloor n/4 \rfloor$ edge guards suffice except for a few small values of n.

UNSOLVED PROBLEM 6   Mirror Walls

For any polygon P whose edges are perfect mirrors, prove (or disprove) that only one guard is needed to cover P. (This problem is often stated in the language of the theory of billiards.) In one variant of the problem, any light ray that directly hits a vertex is absorbed.

The art gallery theorem shows that placing a guard at every vertex of the polygon is three times more than needed to cover it. But what about for a polyhedron in three dimensions? It seems almost obvious that guards at every vertex of any polyhedron should cover the interior of the polyhedron. It is remarkable that this is not so.

The reason the art gallery theorem succeeds in two dimensions is the fundamental property that all polygons can be triangulated. Indeed, Theorem 1.4 is not available to us in three dimensions: not all polyhedra are tetrahedralizable, as demonstrated earlier in Figure 1.7(c). If our polyhedron indeed was tetrahedralizable, then every tetrahedron would have guards in the corners, and all the tetrahedra would then cover the interior.

Exercise 1.41. *Let P be a polyhedron with a tetrahedralization where all edges and diagonals of the tetrahedralization are on the boundary of P. Make a conjecture about the number of guards needed to cover P.*

Exercise 1.42. *Show that even though the Schönhardt polyhedron (Figure 1.7) is not tetrahedralizable, it is still covered by guards at every vertex.*

Because not all polyhedra are tetrahedralizable, the "obviousness" of coverage by guards at vertices is less clear. In 1992, Raimund Seidel constructed a polyhedron such that guards placed at every vertex do *not* cover the interior. Figure 1.17 illustrates a version of the polyhedron. It can be constructed as follows. Start with a large cube and let $\varepsilon \ll 1$ be a very small positive number. On the front side of the cube, create an $n \times n$ array of 1×1 squares, with a separation of $1 + \varepsilon$ between their rows and columns. Create a tunnel into the cube at each square that does not quite go all the way through to the back face of the cube, but instead stops ε short of that back face. The result is a deep dent at each square of the front face. Repeat this procedure for the top face and the right face, staggering the squares so their respective dents do not intersect. Now imagine standing deep in the interior, surrounded by dent faces above and below, left and right, fore and aft. From a sufficiently central point, no vertex can be seen!

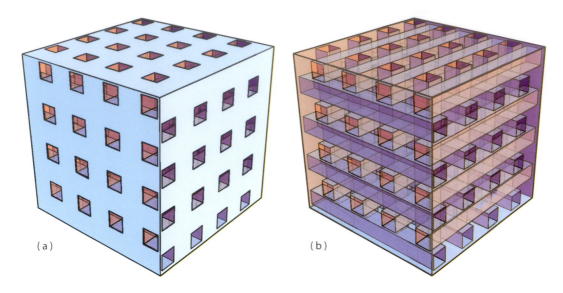

(a)   (b)

Figure 1.17. (a) The Seidel polyhedron with (b) three faces removed to reveal the interior.

Exercise 1.43. *Prove the above claim, which implies that guards at every vertex of the Seidel polyhedron do not cover the entire interior. Notice that this implies the Seidel polyhedron is not tetrahedralizable.*

Exercise 1.44. *Let n be the number of vertices of the Seidel polyhedron. What order of magnitude, as a function of n, is the number of guards needed to cover the entire interior of the polyhedron? (See the Appendix for the Ω notation that captures this notion of "order of magnitude" precisely.)*

1.4 SCISSORS CONGRUENCE IN 2D

The crucial tool we have employed so far is the triangulation of a polygon P by its diagonals. The quantities that have interested us have been combinatorial: the number of edges of P and the number of triangles in a triangulation of P. Now we loosen the restriction of only cutting P along diagonals, permitting arbitrary straight cuts. The focus will move from combinatorial regularity to simply preserving the area.

A *dissection* of a polygon P cuts P into a finite number of smaller polygons. Triangulation can be viewed as an especially constrained form of dissection. The first three diagrams in Figure 1.18 show dissections of a square. Part (d) is not a dissection because one of the partition pieces is not a polygon.

Given a dissection of a polygon P, we can rearrange its smaller polygonal pieces to create a new polygon Q of the same area. We say two

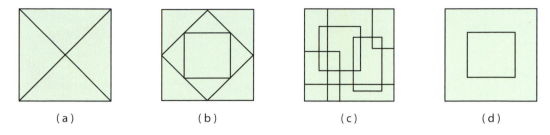

Figure 1.18. Three dissections (a)–(c) of a square, and (d) one that is not a dissection.

polygons P and Q are *scissors congruent* if P can be cut into polygons $P_1, \ldots, P_n$ which then can be reassembled by rotations and translations to obtain Q. Figure 1.19 shows a sequence of steps that dissect the *Greek cross* and rearrange the pieces to form a square, detailed by Henry Dudeney in 1917. However, the idea behind the dissection appears much earlier, in the work of the Persian mathematician and astronomer Mohammad Abu'l-Wafa Al-Buzjani of the tenth century.

The delight of dissections is seeing one familiar shape surprisingly transformed to another, revealing that the second shape is somehow hidden within the first. The novelty and beauty of dissections have attracted puzzle enthusiasts for centuries. Another dissection of the Greek cross, this time rearranged to form an equilateral triangle, discovered by Harry Lindgren in 1961, is shown in Figure 1.20.

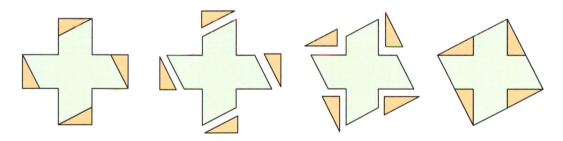

Figure 1.19. The Greek cross is scissors congruent to a square.

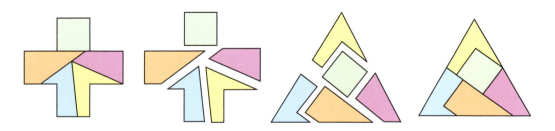

Figure 1.20. Lindgren's dissection of a Greek cross to an equilateral triangle.

Exercise 1.45. *Find another dissection of the Greek cross, something quite different from that of Figure 1.19, that rearranges to form a square.*

Exercise 1.46. *Find a dissection of two Greek crosses whose combined pieces form one square.*

Exercise 1.47. *Show that any triangle can be dissected using at most three cuts and reassembled to form its mirror image. As usual, rotation and translation of the pieces are permitted, but not reflection.*

Exercise 1.48. *Assume no three vertices of a polygon P are collinear. Prove that out of all possible dissections of P into triangles, a triangulation of P will always result in the fewest number of triangles.*

If we are given two polygons P and Q, how do we know whether they are scissors congruent? It is obvious that they must have the same area. What other properties or characteristics must they share? Let's look at some special cases.

Lemma 1.49. *Every triangle is scissors congruent with some rectangle.*

Figure 1.21 illustrates a proof of this lemma. Given any triangle, choose its longest side as its base, of length b. Make a horizontal cut halfway up from the base. From the top vertex, make another cut along the perpendicular from the apex. The pieces can then be rearranged to form a rectangle with half the altitude a of the triangle and the same base b. Note this dissection could serve as a proof that the area of a triangle is $ab/2$.

Lemma 1.50. *Any two rectangles of the same area are scissors congruent.*

Proof. Let R_1 be an $(l_1 \times h_1)$-rectangle and let R_2 be an $(l_2 \times h_2)$-rectangle, where $l_1 \cdot h_1 = l_2 \cdot h_2$. We may assume that the rectangles are not identical, so that $h_1 \neq h_2$. Without loss of generality, assume $h_2 < h_1 \leq l_1 < l_2$.
  We know from $l_1 < l_2$ that rectangle R_2 is longer than R_1. However, for this construction, we do not want it to be *too* long. If $2l_1 < l_2$,

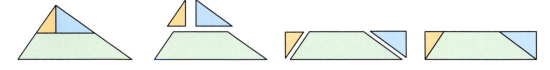

Figure 1.21. Every triangle is scissors congruent with a rectangle.

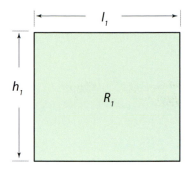
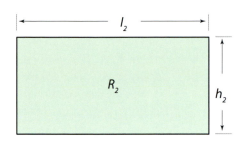

Figure 1.22. Two rectangles satisfying $h_2 < h_1 \leq l_1 < l_2 < 2l_1$.

then cut R_2 in half (with a vertical cut) and stack the two smaller rectangles on one another. This stacking will reduce the length of R_2 by half but will double its height. However, because $l_1 \cdot h_1 = l_2 \cdot h_2$, the height of the stacked rectangles $2h_2$ will still be less than h_1. Repeat this process of cutting and stacking until we have two rectangles with $h_2 < h_1 \leq l_1 < l_2 < 2l_1$, as shown in Figure 1.22.

After placing R_1 and R_2 so that their lower left corners coincide and they are flush along their left and base sides, draw the diagonal from x, the top left corner of R_1, to y, the bottom right corner of R_2. The resulting overlay of lines, as shown in Figure 1.23(a), dissects each rectangle into a small triangle, a large triangle, and a pentagon. We claim that these dissections result in congruent pieces, as depicted in Figure 1.23(b). It is clear the pentagons C are identical. In order to see that the small triangles A_1 and A_2 are congruent, first notice that they are similar to each other as well as similar to the large triangle xoy, as labeled in Figure 1.23(a). Using $l_1 \cdot h_1 = l_2 \cdot h_2$, the equation

$$\frac{h_1 - h_2}{l_2 - l_1} = \frac{h_1}{l_2} \qquad (1.2)$$

can be seen to hold by cross-multiplying. Because A_1 is similar to xoy, whose altitude/base ratio is h_1/l_2, and the height of A_1 is $h_1 - h_2$, equation (1.2) shows that the base of A_1 is $l_2 - l_1$. But since the base

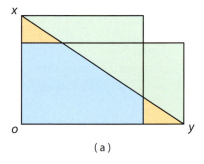
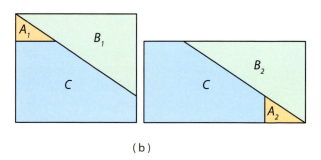

(a)   (b)

Figure 1.23. Any two rectangles of the same area are scissors congruent.

length of A_2 is $l_2 - l_1$, it follows that A_1 and A_2 are congruent. A nearly identical argument shows that the large triangles B_1 and B_2 are congruent. The theorem follows immediately.   □

Exercise 1.51. *Let polygon P_1 be scissors congruent to polygon P_2, and let polygon P_2 be scissors congruent to polygon P_3. Show that polygon P_1 is scissors congruent to polygon P_3. In other words, show that scissors congruence is transitive.*

Exercise 1.52. *Dissect a 2×1 rectangle into three pieces and rearrange them to form a $\sqrt[3]{4} \times \sqrt[3]{2}$ rectangle.*

It is immediate that scissors congruence implies equal area, but the converse is by no means obvious. This fundamental result was proved by Farkas Bolyai in 1832 and independently by Paul Gerwien in 1833.

Theorem 1.53 (Bolyai-Gerwein). *Any two polygons of the same area are scissors congruent.*

Proof. Let P and Q be two polygons of the same area α. Using Theorem 1.4, dissect P into n triangles. By Lemma 1.49, each of these triangles is scissors congruent to a rectangle, which yields n rectangles. From Lemma 1.50, these n rectangles are scissors congruent to n other rectangles with base length 1. Stacking these n rectangles on top of one another yields a rectangle R with base length 1 and height α. Using the same method, we see that Q is scissors congruent with R as well. Since P is scissors congruent with R, and R with Q, we know from Exercise 1.51 that P is scissors congruent with Q.   □

Example 1.54. The Bolyai-Gerwein theorem not only proves the existence of a dissection, it gives an algorithm for constructing a dissection. Consider the Greek cross of Figure 1.19, say with total area 5/2. We give a visual sketch of the dissection implied by the proof of the theorem to show scissors congruence with a square of the same area. The first step is a triangulation, as shown in Figure 1.24, converting the cross into 10 triangles, each of area 1/4 and base length 1. Second, each triangle is dissected to a rectangle of width 1 and height 1/4. Finally these are stacked to form a large rectangle of area 5/2.

Now starting from the square of area 5/2, a triangulation yields two triangles of base length $\sqrt{5}$, as shown in Figure 1.25. Each triangle is then transformed into a $\sqrt{5}/4 \times \sqrt{5}$ rectangle. Each rectangle needs to be transformed into another rectangle of base length 1 (and height 5/4). Since this rectangle is too long (as described in the proof of Lemma 1.50), it needs to be cut into two pieces and stacked. Then, the (stacked) rectangle is cut and rearranged to form two rectangles of base length 1.

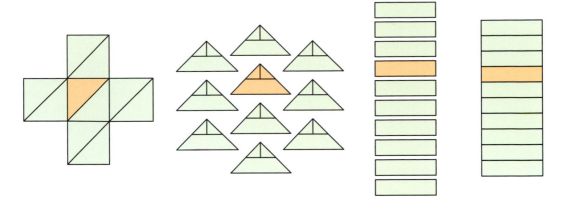

Figure 1.24. Cutting the Greek cross into a rectangle of base length 1 using the Bolyai-Gerwein proof. The transformations to the colored triangle are tracked through the stages.

Although the Bolyai-Gerwein proof is constructive, it is far from optimal in terms of the number of pieces in the dissection. Indeed, we saw in Figure 1.19 that a 5-piece dissection suffices to transform the Greek cross to a square.

Exercise 1.55. *Following the proof of the Bolyai-Gerwein theorem, what is the actual number of polygonal pieces that results from transforming the Greek cross into a square? Assume the total area of the square is 5/2 and use Figures 1.24 and 1.25 for guidance.*

★ **Exercise 1.56.** *Show that a square and a circle are not scissors congruent, even permitting curved cuts.*

It is interesting to note that the Bolyai-Gerwien theorem is true for polygons not only in the Euclidean plane, but in hyperbolic and elliptic geometry as well.

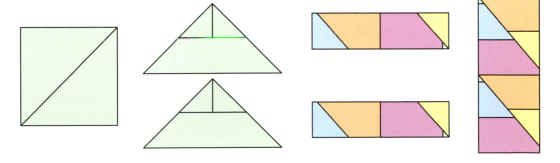

Figure 1.25. Cutting the square into a rectangle of base length 1 using the Bolyai-Gerwein proof. The last transformation is color-coded to show the fit of the pieces.

UNSOLVED PROBLEM 7   **Fair Partitions**

For each positive integer n, is it always possible to partition a given convex polygon into n convex pieces such that each piece has the same area and the same perimeter? This has been established only for $n = 2$ and $n = 3$.

1.5 SCISSORS CONGRUENCE IN 3D

From the discussion above, we see that equal area suffices to guarantee a dissection of one polygon to another. Is this true in higher dimensions? That is, if we are given two polyhedra of the same volume, can we make them scissors congruent? In 1900, in his famous address to the International Congress of Mathematicians, the renowned mathematician David Hilbert asked the same question: Are any two polyhedra of the same volume scissors congruent? This problem was solved in the negative by Hilbert's student Max Dehn a few years later. Indeed, Dehn constructed two tetrahedra with congruent bases and the same height which are *not* scissors congruent. In order to understand Dehn's results, we need to take a closer look at polyhedra.

Unlike polygons, where angles only appear at the vertices, polyhedra have angles along edges as well. The angle along each edge of a polyhedron, formed by its two adjacent faces, is called the dihedral angle.

Definition. The *dihedral angle* θ at the edge e of a polyhedron shared between two faces f_1 and f_2 is the angle between two unit normal vectors n_1 and n_2 to f_1 and f_2, respectively. Thus $n_1 \cdot n_2 = \cos\theta$. By convention, the normal vectors point to the exterior of the polyhedron, and the dihedral angle at e is the interior angle.

For example, the dihedral angle along each edge of a cube is $\pi/2$. For further examples of dihedral angles, we will use Figure 1.26, which shows four tetrahedra embedded inside the cube.

Example 1.57. The tetrahedron on the left in Figure 1.27 repeats Figure 1.26(a) with labels. The dihedral angle along the edges AD, BC, and BD is $\pi/2$, and the edges AB and CD have dihedral angles of $\pi/4$. To find the dihedral angle along edge AC, we look back at the decomposition of the cube in Figure 1.6(c). Here the cube is tetrahedralized into six polyhedra congruent to the polyhedron on the left. Thus the dihedral angle along AC is $\pi/3$.

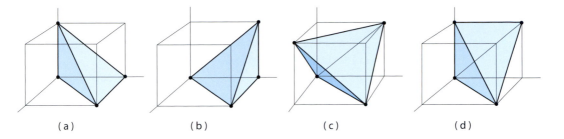

Figure 1.26. Four tetrahedra embedded inside the cube.

Example 1.58. The polyhedron on the right in Figure 1.27 repeats Figure 1.26(b). The dihedral angle along the edges AD, BD, and CD is $\pi/2$, because they are sides of the surrounding cube. Due to symmetry of the polyhedron, the edges AB, AC, and BC have the same dihedral angle. We draw the midpoint E of edge BC in order to calculate the dihedral angle AED along BC. If the cube of Figure 1.26(b) has side length x, then the length of DE is $x/\sqrt{2}$. Because the length of AD is x, the dihedral angle along BC is $\arctan\sqrt{2}$.

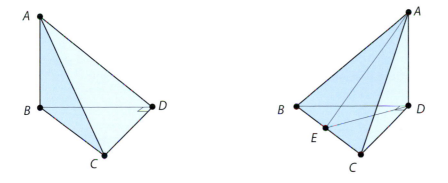

Figure 1.27. Two tetrahedra with congruent bases and the same height.

Exercise 1.59. *Find the dihedral angles of the tetrahedra in Figure 1.26(c) and (d).*

Exercise 1.60. *Find the dihedral angles of a regular dodecahedron.*

The dihedral angles are the key ingredient in understanding the ideas of Dehn. Instead of looking at these angles themselves, Dehn was interested in using them *up to rational multiples of π*. More precisely, let $f : \mathbb{R} \to \mathbb{Q}$ be a function from the real numbers to the rationals that satisfies three properties:

1. $f(v_1 + v_2) = f(v_1) + f(v_2)$ for all $v_1, v_2 \in \mathbb{R}$;
2. $f(qv) = qf(v)$ for all $q \in \mathbb{Q}$ and $v \in \mathbb{R}$;
3. $f(\pi) = 0$.

We call any such function a *d-function* (*d* for dihedral). For instance, for any *d*-function f, we see that

$$f\left(\frac{5\pi}{2}\right) = \frac{5}{2} \cdot f(\pi) = \frac{5}{2} \cdot 0 = 0.$$

Similarly, f maps any rational multiple of π to 0. We define a *rational angle* as an angle that is a rational multiple of π, and an *irrational angle* as one that is not.

For an edge e of a polyhedron, let $l(e)$ denote the length of e and let $\phi(e)$ denote the dihedral angle of e. For any choice of *d*-function f, Dehn's idea is to associate the value

$$l(e) \cdot f(\phi(e))$$

to each edge e, which he called its *mass*. Thus the mass of any edge is 0 when its dihedral angle is rational. We define the *Dehn invariant* of a polyhedron P to be the sum of the masses along the edges of P:

$$D_f(P) := \sum_{e \in P} l(e) \cdot f(\phi(e)).$$

Notice that the Dehn invariant depends on the choice of a *d*-function: For different choices of f, we get different Dehn invariants. The beauty of the Dehn invariant is that it is truly invariant under dissections. This is captured in the following theorem, which we prove using techniques invented by Hugo Hadwiger much later, in 1949.

Theorem 1.61 (Dehn-Hadwiger). *Let f be any d-function. If P is a polyhedron dissected into polyhedra $P_1, P_2, \ldots, P_n$, then*

$$D_f(P) = D_f(P_1) + D_f(P_2) + \cdots + D_f(P_n).$$

Proof. Let f be any *d*-function. The Dehn invariant of P sums the masses of the edges of P. After the dissection of P into several polyhedra, many new edges are introduced. Let e be an edge in the decomposition of P. There are three possible ways for e to appear in P, only one of which contributes to the mass sum.

1. Edge e is contained in an edge of P; see Figure 1.28(a). Let $\phi(e)$ be the dihedral angle of P along e and let $\{\phi_1(e), \phi_2(e), \ldots, \phi_k(e)\}$ be the set of dihedral angles of the polyhedral pieces along e in the decomposition. Then $l(e) \cdot f(\phi_i(e))$ is the mass contributed by the polyhedron P_i along edge e. The sum of the masses along e in the decomposition is

$$l(e) \cdot f(\phi_1(e)) + l(e) \cdot f(\phi_2(e)) + \cdots + l(e) \cdot f(\phi_m(e)),$$

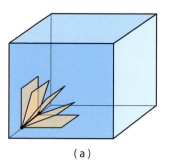 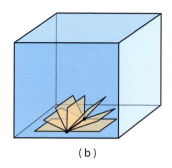 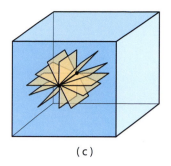

(a)   (b)   (c)

Figure 1.28. An edge of the dissection contained in (a) an original edge, (b) an interior of a face, or (c) the interior of the polyhedron.

which becomes

$$l(e) \cdot f(\phi_1(e) + \phi_2(e) + \cdots + \phi_m(e))$$

by Property 1 of a d-function. Since this is equal to $l(e) \cdot f(\phi(e))$, the masses add up in the required manner.

2. Edge e is contained in the interior of a face of P; see Figure 1.28(b). In this case, the sum of the masses becomes $l(e) \cdot f(\phi(e)) = l(e) \cdot f(\pi) = 0$. So a new edge created from a dissection that appears in the interior of a face of P has no mass.

3. Edge e is contained in the interior of P; see Figure 1.28(c). By a similar argument as before, $l(e) \cdot f(2\pi) = 0$, again contributing no new mass.

Thus the mass sum under the dissection depends only on the edges of P. As each edge e is covered exactly once by dissection edges, whose lengths sum to $l(e)$, the mass sum for any dissection is exactly the same as the mass sum for the original P.   □

Corollary 1.62. *Let P and Q be polyhedra and let f be any d-function. If $D_f(P)$ does not equal $D_f(Q)$, then P and Q are not scissors congruent.*

Proof. We prove this by contradiction. Let P and Q be scissors congruent. Then there is a dissection of P into polyhedra $P_1, P_2, \ldots, P_n$. By the Dehn-Hadwiger theorem,

$$D_f(P) = D_f(P_1) + D_f(P_2) + \cdots + D_f(P_n) = D_f(Q),$$

the last equality following because the rearrangement of the polyhedral pieces forms Q.   □

Example 1.63. Consider the tetrahedron on the left of Figure 1.27, calling it T_1. Example 1.57 shows that its set of dihedral angles is $\{\pi/2, \pi/3, \pi/4\}$. Because all the dihedral angles are rational, the mass for all the edges is 0. Thus $D_f(T_1) = 0$ for any d-function f.

Example 1.64. Consider the tetrahedron on the right of Figure 1.27, calling it T_2. If the cube of Figure 1.26(b) has side length 1, Example 1.58 shows three edges of length 1 with dihedral angle $\pi/2$ and three edges of length $\sqrt{2}$ with dihedral angle $\arctan \sqrt{2}$. Hence,

$$D_f(T_2) = 3\ f\left(\frac{\pi}{2}\right) + 3\sqrt{2}\ f(\arctan \sqrt{2}).$$

The first term is zero, but the second term need not be. For example, f could be the identity function on irrational multiples of π, in which case $D_f(T_2) = 3\sqrt{2} \arctan \sqrt{2} \neq 0$.

These two examples show the tetrahedra T_1 and T_2 in Figure 1.27, both having congruent bases and the same height and so the same volume, have different values for their Dehn invariant. By the Dehn-Hadwiger theorem, T_1 and T_2 are not scissors congruent. Generalizing this example, any polyhedron with all rational dihedral angles can never be dissected to a polyhedron with at least one irrational dihedral angle.

Exercise 1.65. *Show that a regular tetrahedron cannot be scissors congruent with a cube.*

★ **Exercise 1.66.** *Show that no Platonic solid is scissors congruent to any other Platonic solid.*

Although the Dehn-Hadwiger theorem can be used to show that two polyhedra are *not* scissors congruent, it does not directly tell us anything about the converse. In 1965, Jean-Pierre Sydler showed the following to be true, although its proof is quite involved.

Theorem 1.67 (Sydler). *Polyhedra P and Q are scissors congruent if $D_f(P) = D_f(Q)$ for every d-function f.*

This, along with the Dehn-Hadwiger theorem, show that Dehn invariants are a *complete* set of scissors-congruent invariants for polyhedra. To understand the power of this result, consider the tetrahedron in Figure 1.26(a). By Example 1.63, this tetrahedron as well as any cube have D_f equal to zero for any d-function f. Thus, by Sydler's theorem, a dissection of a cube exists whose rearrangement yields this tetrahedron! Indeed, Sydler demonstrated a beautiful construction showing how this dissection works. Figure 1.29 shows the tetrahedron being transformed into a prism whose base is an isosceles right triangle. It is then not hard to imagine how this prism can be made into a rectangular block, which then can be made into a cube.

Exercise 1.68. *Complete the construction above, dissecting the tetrahedron of Figure 1.29 into a cube.*

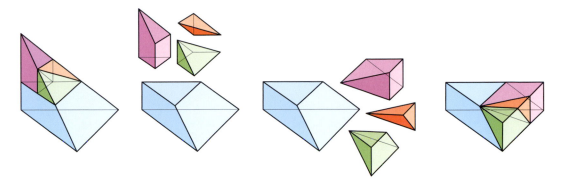

Figure 1.29. Sydler's dissection of a tetrahedron into a prism.

UNSOLVED PROBLEM 8   Dehn Construction

Given two equal-volume polyhedra with all rational dihedral angles, construct an efficient algorithm for finding a dissection, as guaranteed by Sydler's theorem.

★ **Exercise 1.69.** *Let P be a 2 × 1 × 1 rectangular prism. Cut P into eight or fewer pieces and rearrange the pieces to form a cube.*

UNSOLVED PROBLEM 9   Five-Piece Puzzle

Can a 2 × 1 × 1 rectangular prism be cut into five or fewer pieces, which can then be rearranged to form a cube?

SUGGESTED READINGS

Joseph O'Rourke. *Computational Geometry in C*. Cambridge University Press, 2nd edition, 1998.

Chapter 1 of this text covers the first three sections of this chapter with a more algorithmic slant.

Richard Stanley. *Enumerative Combinatorics*, Volume I and II. Cambridge University Press, 1997, 1999.

This monumental two-volume work covers the art of counting combinatorial objects. The classical text in graduate school. In particular, more than 100 objects are described that are all counted by the Catalan number.

Joseph O'Rourke. *Art Gallery Theorems and Algorithms*. Oxford University Press, 1987.

A monograph on art gallery theorems, now more than 20 years old, but still (we think) the first source to consult. Out of print but available at `http://cs.smith.edu/~orourke//books/ArtGalleryTheorems/art.html`. For a more recent survey, see Jorge Urrutia, "Art gallery and illumination problems" (in Jörg-Rüdiger Sack and Jorge Urrutia, editors, *Handbook of Computational Geometry*, chapter 22, pages 973–1027, Elsevier, 2000).

Greg Frederickson. *Dissections: Plane & Fancy*. Cambridge University Press, 1997.

A delightfully readable book on dissections of polygons and polyhedra. Frederickson also wrote two other books on more specialized dissections: *Hinged Dissections: Swinging & Twisting* (Cambridge University Press, 2002) and *Piano-Hinged Dissections: Time to Fold!* (A K Peters, 2006).

Vladimir Boltyanskii, *Hilbert's Third Problem*. V. H. Winston & Sons, 1978.

A gem of a book, covering a broad scope of problems and proofs related to scissors congruence in 2D and 3D. Includes a reworking of Sylder's proof of the converse of the Dehn-Hadwiger theorem.

CONVEX HULLS **2**

The next three chapters focus on finite *point sets*, discrete structures as fundamental as polygons for computational geometry. Unlike polygon vertices, point sets are unordered. Often the first step in organizing a point set is finding its convex hull, the topic of this chapter. We start with the foundations of convexity (Section 2.1) and then provide a first algorithm, the incremental algorithm (Section 2.2). After a brief detour introducing the art of algorithm analysis (Section 2.3), we present two more algorithms: gift wrapping and the Graham scan (Section 2.4). A lower bound (Section 2.5) then shows that in some sense the algorithm search is over, because the Graham scan is "asymptotically optimal." Nevertheless, we push to a fourth algorithm, divide-and-conquer (Section 2.6), not only because it illustrates an important algorithm paradigm, but because it is the only one of the four that generalizes to an optimal algorithm for finding the convex hull in three dimensions (Section 2.7).

2.1 CONVEXITY

In the previous chapter we discussed convex *polygons*. Now we extend the idea to convex *regions*: a region is convex if any two points of the region are visible to one another within the region. In this chapter, given a set of distinct points S, we are interested in constructing its *convex hull*. We can visualize the convex hull intuitively: if each point of S is a nail pounded into the plane, the convex hull is the region enclosed by an elastic rubber band stretched around all the nails. Figure 2.1(a) shows a nonconvex region containing a point set S; part (b) shows a convex region enclosing S. The convex hull of S is given in Figure 2.1(c). There are numerous practical applications for finding the convex hull, including collision detection, Geographic Information Systems (GIS), and pattern recognition.

Based on the intuition provided by the rubber-band analogy, the convex hull is the smallest convex region containing the point set S. The notion of "smallest" can be formally captured as follows.

Definition. The *convex hull* of S, denoted by conv(S), is the intersection of all convex regions that contain S.

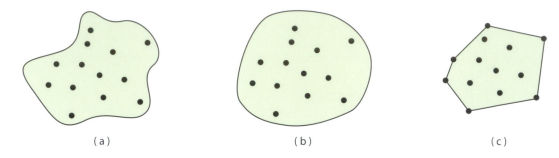

Figure 2.1. A point set S along with (a) a nonconvex region enclosing S, (b) a convex region enclosing S, and (c) the convex hull of S.

Exercise 2.1. *Show that the use of the word "convex" in convex hull is justified; that is, show that* conv(S) *is indeed a convex region.*

Although this definition of convex hull is natural, it is not computationally useful in that it does not immediately suggest a way actually to *find* the convex hull of a point set. Intersecting *all* the convex sets containing S is hardly an option. An alternative characterization based on visibility will eventually lead us to a computational method of finding the convex hull.

Recall that a region is convex if and only if any two points of the region are visible to each other. In other words, for any two points x and y in a convex region R, the line segment xy is also in R. Any point of the line segment xy can be written as $\alpha x + \beta y$, where $\alpha \geq 0$, $\beta \geq 0$, and $\alpha + \beta = 1$. So as α varies from 1 to 0, β varies in lock-step from 0 to 1, and the point moves along the segment xy, starting at x and ending at y.

This idea can be generalized to an arbitrary number of points: A *convex combination* of points $S = \{p_1, \ldots, p_n\}$ is of the form

$$\lambda_1 p_1 + \cdots + \lambda_n p_n,$$

where $\lambda_i \geq 0$ and $\sum \lambda_i = 1$. In some sense, the set of convex combinations tries to capture visibility from all points in the set S. The following theorem shows that this is exactly what we seek.

Theorem 2.2. *For a point set $S = \{p_1, \ldots, p_n\}$, the convex hull of S is the set of all convex combinations of S.*

Proof. Let M be the set of convex combinations of S. Formally,

$$M = \left\{ \lambda_1 p_1 + \cdots + \lambda_n p_n \,\middle|\, \lambda_i \geq 0, \sum \lambda_i = 1 \right\}.$$

In order to prove conv$(S) = M$, we show conv$(S) \subseteq M$ and $M \subseteq$ conv(S).

Part I: $conv(S) \subseteq M$. It is easy to see that M contains S. If $\lambda_i = 1$ and all others are 0, we get p_i in M. Thus, if we can prove that M is a convex region, then $conv(S) \subseteq M$ since $conv(S)$ is the intersection of all convex regions containing S. Let x and y be any two points in M; we need to verify the segment xy is in M. Since x is in M, it can be written as

$$x = \lambda_1 p_1 + \cdots + \lambda_n p_n , \tag{2.1}$$

where $\lambda_i \geq 0$ and $\sum \lambda_i = 1$. Similarly, y can be written as

$$y = \lambda'_1 p_1 + \cdots + \lambda'_n p_n,$$

where $\lambda'_i \geq 0$ and $\sum \lambda'_i = 1$. Moreover, any point of the segment xy can be expressed as

$$\alpha x + \beta y = \alpha \sum \lambda_i p_i + \beta \sum \lambda'_i p_i = \sum (\alpha \lambda_i + \beta \lambda'_i) p_i,$$

for $\alpha \geq 0$, $\beta \geq 0$, and $\alpha + \beta = 1$. Because $(\alpha \lambda_i + \beta \lambda'_i) \geq 0$, and

$$\sum (\alpha \lambda_i + \beta \lambda'_i) = \alpha \sum \lambda_i + \beta \sum \lambda'_i = \alpha \cdot 1 + \beta \cdot 1 = 1,$$

it follows that the segment xy is in M. Thus M is a convex region.

Part II: $M \subseteq conv(S)$. We show that any point in M, which may be expressed as in equation (2.1), is in $conv(S)$ by induction on n. It is clear this is true when $n = 1$; then $M = conv(S) = p_1$. Assume it is true for every point set S' containing fewer than n points, and now consider the set S with n points, $p_1, \ldots, p_n$. By the induction hypothesis, any point

$$x = \lambda'_1 p_1 + \cdots + \lambda'_{n-1} p_{n-1}$$

is in $conv(S') \subset conv(S)$ if and only if $\lambda'_i \geq 0$ and $\sum_i \lambda'_i = 1$. Now we choose $\lambda'_i = \lambda_i / (1 - \lambda_n)$ because $\lambda_1 + \cdots + \lambda_{n-1} = 1 - \lambda_n$. Note that we still satisfy the conditions above. And because $conv(S') \subset conv(S)$, we know that x is in $conv(S)$. Because p_n and x are both in $conv(S)$, and since $conv(S)$ is convex, then any point in the segment xp_n is in $conv(S)$. Thus

$$(1 - \lambda_n) \left(\frac{\lambda_1}{1 - \lambda_n} p_1 + \cdots + \frac{\lambda_{n-1}}{1 - \lambda_n} p_{n-1} \right) + \lambda_n p_n = \lambda_1 p_1 + \cdots + \lambda_n p_n$$

is in $conv(S)$, where $\sum \lambda_i = 1$. $\qquad \square$

This theorem gives considerable insight into what constitutes the convex hull, but it is not yet algorithmic. We turn to our first algorithm in the next section.

Exercise 2.3. *Let S be the four points $\{(0, 0), (0, 1), (1, 0), (1, 1)\}$ in the plane. Show using Theorem 2.2 that $conv(S)$ is the square with vertices at S.*

Exercise 2.4. For a point set S in the plane with at least four points, show that S can be partitioned into two sets A and B such that $\text{conv}(A)$ intersects $\text{conv}(B)$.

Exercise 2.5. Show that $\text{conv}(S)$ is the convex polygon with the smallest perimeter that contains S.

Exercise 2.6. Show that $\text{conv}(S)$ is the convex polygon with the smallest area containing S.

★ *Exercise 2.7.* Show that the rectangle of smallest area enclosing a point set S has at least one of its sides flush with (i.e., containing) an edge of $\text{conv}(S)$.

2.2 THE INCREMENTAL ALGORITHM

Given a point set S in the plane, how do we compute the convex hull? In fact, what does it even mean to *compute* the convex hull? A natural representation is the boundary of the polygon $\text{conv}(S)$, called the *hull* of the point set S. The vertices on the hull are referred to as *hull vertices*. Computing the convex hull means identifying the hull vertices.[1]

If someone were to give you a piece of paper with a marked set of points S, it is easy for your eyes to find the hull of S. But what if you were given a collection of points that were listed using their coordinates (x, y) in the plane, as in Figure 2.2, which is how data are usually stored? Naturally, the extreme points are in the hull, the leftmost and rightmost points (extreme in x) and the highest and lowest points (extreme in y). It is more difficult to identify the other hull points. Figure 2.2 shows a data set and the corresponding points in the plane. Finding the hull points (colored) in the data set without graphing is not easy, even with just the 18 points considered. For point sets in the thousands or more, visualization breaks down.

Given a point set as a list expressed in coordinates, our goal will be to identify the hull points. The creativity needed to construct a proof of a theorem is not identical to that needed to construct an *algorithm*, a blueprint of instructions on how to construct what the theorem ensures exists. As our emphasis is on the geometry that underlies both proof and algorithms, we will proceed through a series of algorithms for the convex hull, stressing geometric intuition throughout.

---

[1] The term *convex hull* usually means the convex set $\text{conv}(S)$ in the mathematics literature, and the convex polygonal boundary in the computer science literature. Here we opt for the former meaning of *convex hull* and (somewhat nonstandardly) use *hull* for its boundary.

(2, 10)	(10, 17)
(3, 3)	(10, 13)
(4, 15)	(12, 5)
(5, 7)	(12, 18)
(6, 11)	(13, 3)
(7, 2)	(13, 13)
(7, 16)	(15, 8)
(8, 5)	(15, 15)
(9, 9)	(17, 11)

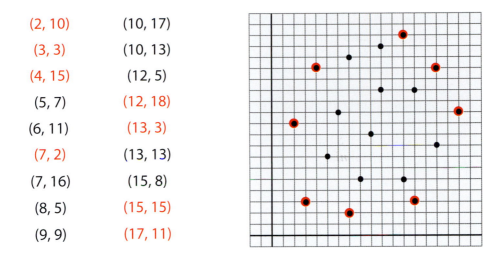

Figure 2.2. The left side lists 18 points, with the hull points colored red. The right side shows the plot of these points in the plane, again with the hull points marked.

We begin with an algorithm that is closest to a mathematical proof, namely proof by induction. At a high level, we assume the hull of k points has been constructed and use this to build the hull with $k + 1$ points. This is called the *incremental algorithm*; Figure 2.3 shows this algorithm in action.

The foundation of this algorithm is based on an ordering of our point set S. We order the points of S based on their x-coordinates. Note that if two or more points share the same x-coordinate, we can always rotate the plane slightly such that all points have distinct x-coordinates. So we

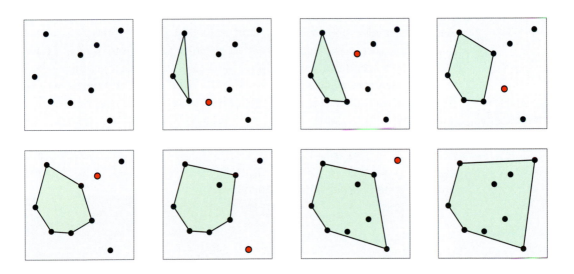

Figure 2.3. The incremental algorithm in action.

assume for simplicity of exposition that the x-ordering is unique, and will return to this issue later in the next section.

Let H_3 be the first three points in S ordered such that they traverse counterclockwise around the triangle conv(H_3). Now assume H_k is the set of hull points of the first k points in S, which is ordered counterclockwise around conv(H_k). Assuming we have constructed H_k, consider the next point p on our list S. Since this list is ordered, p belongs to H_{k+1} because it is extreme in the x-direction. But by adding p to our hull, several of the previous hull points might become interior to the polygon. So we now need a way to add p to our hull list H_k in the appropriate position while removing possible extraneous points.

Definition. Let P be a convex polygon and x a point on the boundary of P. Then a line L containing x *supports* P at x if all of P lies on one side of L. Line L is then called a *tangent* to P at x.

Our task is to find two points in H_k which have tangent lines to conv(H_k) passing through p. Figure 2.4 shows an example of (a) the convex hull conv(H_k) with a point p, (b) two points in H_k with the desired tangent lines through p, and (c) the new conv(H_{k+1}). But how can we find these two special tangency points on H_k?

The key observation is to see things from the perspective not of the vertices of conv(H_k), but of its edges. As Figure 2.4(b) shows, each edge of conv(H_k) is either visible to p, invisible to p, or lies on the same line as p. To keep matters simple, we skirt this last possibility by assuming that no three points of S lie on a line. So there will be two vertices of the convex hull where edges switch from being visible to being invisible from p — these are the tangency vertices. Consider any edge of the polygon conv(H_k), and let L denote the line on which this edge lies. Then observe that an edge is visible to p if p and conv(H_k) lie on opposite sides of L. Similarly, the edge is invisible if they lie on the same side of L; see Figure 2.5. So the points we seek are those with tangent lines that group and separate conv(H_k) and p.

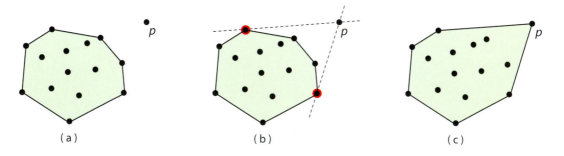

Figure 2.4. Convex hull of k points and the incremental addition of another point.

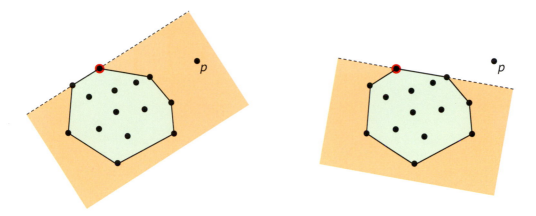

Figure 2.5. Tangent line lines transitioning from grouping and separating the convex hull from the new point p.

Once the two points of tangency are found, we simply insert them into the appropriate position in conv(H_k), removing the new interior points. Thus, if p_i and p_j are the special vertices of tangency, where p_i appears before p_j in the counterclockwise ordering of H_k, then $H_{k+1} = \{\ldots, p_{i-1}, p_i, p, p_j, p_{j+1}, \ldots\}$.

Exercise 2.8. *Argue that the incremental algorithm terminates.*

Exercise 2.9. *Let the* diameter *of a point set S be the largest distance between any two points of S. Prove that the diameter of S is achieved by two hull vertices.*

★ **Exercise 2.10.** *Prove that if S is the set of n points sampled from a uniform distribution in a unit square, then the expected number of points on the hull of S is of order $O(\log n)$.*

2.3 ANALYSIS OF ALGORITHMS

Although the incremental algorithm uses the geometry of visibility to construct the convex hull, it does not take long to realize that, as described, it is not very efficient. For instance, as shown in Figure 2.3, even when the actual hull of the point set has only a handful of points, intermediate stages of the algorithm repeatedly process points that ultimately may be discarded. It is natural to seek better algorithms. Before moving on, however, there are two concerns that need to be addressed.

1. What does it mean for one algorithm to be *better* than another?
2. Does our algorithm work for all types of point sets in the plane?

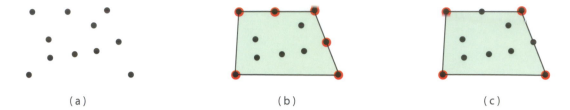

(a)   (b)   (c)

Figure 2.6. A point set along with hull points and extreme points.

Let's start with the second question first. At first glance, the incremental algorithm seems to succeed for any point set in the plane. When we look closer, we realize that not all cases have been covered. Indeed, we have ignored certain special cases that can occur, such as having points with the same x-coordinates, or three collinear points. We said the former could be relieved by slight rotation in the plane, but the latter can only be avoided either by perturbing the point positions (which may alter the hull) or by stipulating that our input set S simply should not have such troublesome triples. A set of points (or more general geometric objects) are said to be in *general position* (i.e., generic position) if they avoid the troublesome configurations, known as *degenerate* situations.

Consider the point set illustrated in Figure 2.6(a). Here even the notion of the hull is ambiguous. Should the hull include all the vertices of the boundary, even flat vertices as in (b), or just the extreme corners, shown in (c)? What constitutes "general position" depends on the algorithm. For the incremental algorithm, we saw two degeneracies: two or more points on a vertical line, and three or more points on any line. For some algorithms we consider in the next two chapters, four points on a circle constitute a degeneracy. In general, degenerate cases arise when there is some algebraic relationship between the points that plays a role in the algorithm.

We will often assume general position for our algorithm inputs to permit us to focus on the essential geometric ideas. But the reader should be aware that any programmed implementation of an algorithm must deal with degeneracies, either handling them correctly or excluding them by fiat. We take some comfort in knowing that a point set is in general position with probability 1 if the n points are chosen randomly. But the flip side of this coin is that inputs derived from the real world are rarely random. The Suggested Readings at the end of this chapter point to resources on this important topic.

Exercise 2.11. *Find a way to order points in the plane without moving them into general position.*

Exercise 2.12. *Alter the incremental algorithm so that it still works for point sets which may have two or more points with the same x-coordinate, without rotating the set into general position.*

Exercise 2.13. *Adjust the incremental algorithm so that it still works for point sets that may include three or more collinear points.*

Now let's address our first question: what does it mean to find a *better* algorithm than the incremental one? One measurement of quality is *how fast* an algorithm can solve the problem. Here the notion of speed is with respect not to some particular computer standard, but to the *number of steps* needed to finish the algorithm. Of course, the number of steps needed depends not just on the algorithm but on the given point set. Moreover, what constitutes a "step" is not evident. Complexity analysis captures the speed of an algorithm by expressing the growth rate of its running time with respect to the input size, using the *big-Oh* notation. The reader unfamiliar with this concept should consult the Appendix, which gives a thumbnail sketch of the main ideas. We repeat a few reminders here.

For an input of size n (such as n numbers or n point coordinates), a running time of $O(f(n))$, where $f(n)$ is some function of n, means that $cf(n)$ is an *upper bound* on the running time of the algorithm, for some constant $c > 0$ and for sufficiently large n. The phrase "for sufficiently large n" implies that only the eventual asymptotic behavior is of interest and allows ignoring all but the dominant terms of the function $f(n)$. By just seeking an upper bound, a bound that holds even for the worst-case input, we remove the dependence of the running time on any characteristics of the input.

Any computation that has running time $O(1)$ is a *constant-time* computation and a "step" of an algorithm is any such computation. For example, one step of the incremental algorithm is to determine whether or not an edge ab of $\text{conv}(H_k)$ is visible to p. This can be accomplished by determining whether (a, b, p) forms a clockwise or counterclockwise triple. Ignoring the details of this computation, it is clear it does not depend on n, the number of points in S. Thus it is an $O(1)$ time computation.

A key complexity result is that sorting n numbers can be accomplished in $O(n \log n)$ time, an upper bound result holding for worst-case input. Moreover, for this problem, $n \log n$ has also been established as a *lower bound*. (This is an intricate result as it must cover all conceivable algorithms, and even then it only holds in certain "models of computation," such as the *decision-tree* model. See the Appendix for further details.) The notation used to express this lower bound is $\Omega(n \log n)$. Thus, having identical upper and lower bounds firmly nails the computational complexity of sorting algorithms.

With this background, let's compute the running time of the incremental algorithm described above. Given n points in S, we first sort them by their x-coordinate. This can be accomplished in $O(n \log n)$ time. For each point of S (after the first three), we need to test each edge of the current hull to see whether it is visible to the point. In the worst case, we might need to consider $k - 1$ edges when adding the kth point. Since this test needs to be executed for each new point, we could perform as many as

$$3 + 4 + \cdots + (n - 1) = \frac{n(n-1)}{2} - (1 + 2) = \frac{n^2}{2} - \frac{n}{2} - 3$$

constant-time computations. Since the most significant term of this sum is n^2, the time complexity is $O(n^2)$. Note that the sorting step with $O(n \log n)$ time is dominated by and therefore absorbed into this $O(n^2)$ time. Thus the incremental algorithm as described has *quadratic* time complexity. We capture the incremental algorithm at a high level as follows:

INCREMENTAL   Convex Hull Algorithm $O(n^2)$

Sort the points of S according to their x-coordinate. The first three of these points determine a triangle, our starting hull. Consider the next point in the ordered set S, add it to the hull, and remove the enclosed non-hull points. Continue this process of adding one point of S at a time until all of S has been processed.

Exercise 2.14. *Given three points (a, b, c), detail a computation that will decide if they form a clockwise or a counterclockwise triangle.*

★ **Exercise 2.15.** *In the incremental algorithm, find a method to search for the tangent lines that leads, overall, to a time complexity of $O(n)$ rather than $O(n^2)$. Notice that this improves the speed of the algorithm to $O(n \log n)$.*

2.4 GIFT WRAPPING AND GRAHAM SCAN

The incremental algorithm followed the example of a standard mathematical proof by induction. Although a time complexity of $O(n^2)$ might seem acceptable, if n is a million, n^2 is a trillion, which leads to unacceptable computation times even with a small time per step. The intermediate stages of the incremental algorithm potentially process all the points of S, rather than trying to find the hull points more directly. Is there a way to find the hull points more quickly? One approach is to wrap a string around the entire point set, where the string catches and turns at each of the hull points. This method is appropriately called the *gift-wrapping* algorithm, first suggested by Donald Chand and Sham Kapur in 1970 (primarily for hulls in higher dimensions). Figure 2.7 shows this algorithm in action.

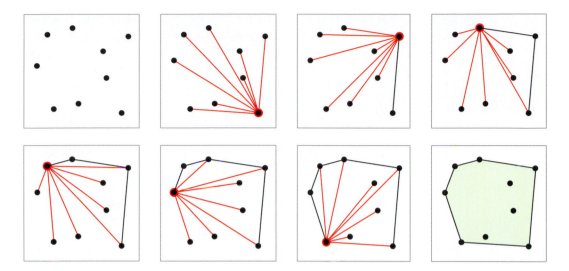

Figure 2.7. The gift-wrapping algorithm in action.

Let's look at the details. Let S be a point set in general position, with no three points collinear. Consider the bottommost point; if there are ties, choose the rightmost. This point will be the anchor from which we wind our string around S. From this point, draw a line segment to all other points in S. Choose the next anchor point on our wrapping of S to be the point that makes the largest angle with the horizontal line. Now that we have started our wrapping, we continue as follows. From the new anchor, draw a line segment to all other points in S, and choose the next anchor forming the largest angle with the last constructed hull edge. This is repeated until the algorithm ends, winding around the entire hull.

Exercise 2.16. *Prove that the point forming the largest angle to the previous edge must be a hull point.*

Exercise 2.17. *Show that this algorithm indeed produces the convex hull, closing up the polygon at the starting point.*

Exercise 2.18. *We phrased the gift-wrapping algorithm in terms of angle comparisons, which are notoriously slow and numerically unstable when implemented naively. Show that angle comparisons can be replaced by* LEFT-OF *tests, where* LEFT-OF(a, b, c) *is true exactly when c is left of the directed line through a and b.*

What about the time complexity of the gift-wrapping algorithm? At each point, the angle to all other points must be calculated, which is n. This needs to be done as many times as there are points on the hull. Thus for S with n points and h hull points, the time complexity of the gift-wrapping algorithm is $O(nh)$. At worst, h could be n, which renders gift wrapping the same worst-case time complexity as the incremental algorithm. But this algorithm has the advantage of being *output-sensitive*: the complexity is proportional to the final hull size h. We summarize as follows:

GIFTWRAPPING   Convex Hull Algorithm $O(nh)$

Start with a known point on the hull as an anchor, such as the bottommost point. Comparing angles to all other points from this anchor, choose the point with the largest angle. Repeat this process, moving around the hull, analogous to the process of winding a string around the point set.

Exercise 2.19. *Describe a point set with n points that serves as the worst-case for the gift-wrapping algorithm.*

Exercise 2.20. *Describe a point set with n points that constitutes the best-case for the gift-wrapping algorithm. What is its time complexity in this case?*

At Bell Laboratories in the late 1960s, Ron Graham needed an algorithm to compute the convex hull of about $10,000$ points in the plane. The $O(n^2)$-time algorithm was too slow for this practical use on the processors of that period, so Graham developed a simple algorithm to meet this need. His 1972 paper is arguably the first publication in computational geometry. Instead of calculating the angle at each point on the hull, Graham's insight was to initially *sort* the points based on these angles. Then, with only a little additional work, he was able to eliminate any extraneous points and focus on just the hull points. We now describe this *Graham scan* algorithm, tracking the example in Figure 2.8.

For a point set S in general position, just as in the gift-wrapping algorithm, choose the rightmost bottom point. Sort the remaining points of S by the angle they form with the horizontal line, from the largest angle to the smallest. From here on, the points are processed in this angularly sorted order. For each point c, a calculation is made to determine whether the two endpoints of the last constructed hull edge ab and c form a *left turn* or a *right turn*. If abc is a right turn, this means that b is not part of the hull and so should be removed from consideration (such as points $a = 3$, $b = 4$, and $c = 5$ in Figure 2.8). This discarding continues as long as the last three points form a right turn. We move to the next point in

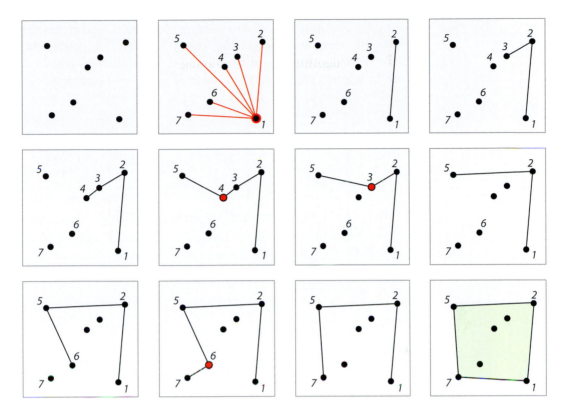

Figure 2.8. The Graham scan algorithm in action. Red points form right turns and are discarded from the hull.

our list once *abc* is a left turn. This continues until we eventually return to the starting point, which completes the hull.

Exercise 2.21. *Alter the Graham scan algorithm so that it still works for points in degenerate position.*

Exercise 2.22. *Describe a point set with n points that is the worst-case for the Graham scan algorithm.*

What is the time complexity of the Graham scan algorithm? As mentioned in Section 2.3, sorting the points has time complexity $O(n \log n)$. As we walk through the algorithm, we see that each point is considered at most twice, once when added and once if it forms an illegal right turn. Since right-turn discarded points are never revisited, the search for hull points executes $O(n)$ iterations. So the overall complexity is $O(n \log n)$, dominated not by the hull search but by the initial sorting of the points.

> **GRAHAM SCAN** Convex Hull Algorithm $O(n \log n)$
>
> Choosing the bottommost point as an anchor, order all other points based on the angles they form about this anchor. Construct the hull by following this ordering, adding points for left hull turns and deleting for right turns.

Exercise 2.23. *Given a point set S, design an algorithm that finds some polygon (any polygon) whose vertices are precisely S.*

Exercise 2.24. *Design a Graham-like algorithm that first sorts the points by their x-coordinate, and then computes the "upper" and "lower" hulls separately. The upper hull is found by repeatedly removing local minima, and the lower hull similarly. Detail the steps needed to realize this high-level description.*

★ **Exercise 2.25.** *Design an algorithm to find the convex hull of a polygon in O(n) time.*

2.5 LOWER BOUND

So far, we have described three algorithms for constructing the convex hull, each superior to the previous in terms of time complexity. All three algorithms are strongly rooted in geometric intuition. The key question now is whether we can improve on the $O(n \log n)$ time complexity achieved by Graham's algorithm. Is there some other geometric insight about point sets that we can use to push the envelope? It turns out the answer is NO, as shown by this beautiful theorem.

Theorem 2.26. *Let S be a point set in the plane. An algorithm that finds the hull points of S in the order of walking around the convex hull cannot be faster than $O(n \log n)$ time. That is, $\Omega(n \log n)$ is a lower bound on the number of comparisons made by a sorting algorithm.*

Proof. We already mentioned that it has been established that $\Omega(n \log n)$ is a lower bound on sorting n numbers in the decision-tree model, which counts comparisons between the numbers being sorted. We now show the same lower bound holds for convex hull algorithms, because if we could construct the hull faster, we could then sort faster.

Suppose we are given an unsorted list of positive numbers $\{x_1, x_2, \ldots, x_n\}$. Construct the set of points in the plane (x_i, x_i^2) as shown in Figure 2.9. Notice that these points lie on the parabola $y = x^2$. We use a convex hull algorithm to construct the hull points.

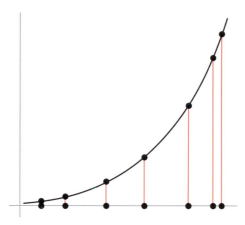 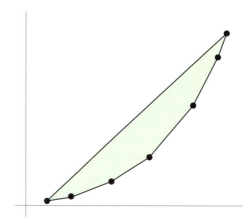

Figure 2.9. Giving height x_i^2 to points x_i in a list, landing on a parabola $y = x^2$.

Because parabolas are convex, every point (x_i, x_i^2) is on the hull. Find the lowest point on the hull in $O(n)$ time. The order in which the points occur on the hull counterclockwise from the lowest point is the sorted order of the input x_i numbers. So we have shown how to use a convex hull algorithm to sort.   □

Incidentally, this is the typical approach to establishing a lower bound on all algorithms that solve a particular problem: reduce the problem to another whose lower bound had been previously established. How to establish lower bounds on some base problems (such as sorting) is a topic for an Algorithms course.

Notice that all the convex hull algorithms we have looked at so far have kept track of the hull points in the order they occur around the convex hull. Theorem 2.26 tells us we cannot find the hull points *in order* any faster. But what if we just wanted to find the hull points without having to worry about their order around the boundary? Can we do things faster if just the *set* of hull points is what we want, rather than an *ordered* set? This was an open problem for many years until 1985, when Franco Preparata and Michael Shamos proved the following:

Theorem 2.27. *A lower bound for any algorithm that identifies the hull points of a point set in the plane is $\Omega(n \log n)$.*

Based on the theorems above, there are no better algorithms (in terms of asymptotic worst-case complexity) for computing the hull of a point set in the plane. But what about point sets in three dimensions? The foundational definitions about convex hulls discussed earlier in this chapter were couched in a general setting, true in all dimensions. Thus the convex hull of points in 3D will yield a convex *polyhedron*. However, all our algorithms for actually calculating the hull points have been focused on point sets in the plane.

Consider the fastest of our algorithms, the Graham scan. Would it be possible to extend this to three dimensions? Remember that the strength of this algorithm comes from being able to sort and order the points based on angles. We were then able to wind around the points in a cyclic manner, ending where we started. No one has found a way to extend the angular scan naturally to 3D. In the following section, we present yet another algorithm, called *divide and conquer*, for computing the hull of points in the plane. The beauty of this algorithm is not just its speed (it is just as fast as the Graham scan) but that it naturally generalizes to three dimensions while achieving the same optimal $O(n \log n)$ time complexity.

Exercise 2.28. *Let S be a set of n points in the plane. Construct an algorithm which finds the convex hull of 3 points of S having the smallest perimeter. What if we want the largest perimeter with 3 points?*

★ **Exercise 2.29.** *Generalize the above approach to finding the convex hull of k points of S having the smallest perimeter. How efficiently can this be solved?*

★ **Exercise 2.30.** *A line is a* best fit *for a point set S in the plane if it mini-mizes the sum of the distances between the points in S and the line. Assuming a convex hull algorithm is available, find the best fit line for a given point set S in the plane.*

2.6 DIVIDE-AND-CONQUER

Whereas the incremental algorithm uses *induction*, the divide-and-conquer algorithm uses the technique of *recursion*, a powerful algorithm paradigm. At a high level, a recursive method divides a problem into smaller problems of the same type. So in order to perform a task, the algorithm calls itself to solve part of the task. We have already used a recursive algorithm earlier, showing in Theorem 1.4 that every polygon has a triangulation. Recall that we did this by taking a polygon, cutting it into two smaller polygons P_1 and P_2 using a diagonal, and then recur-sively using this theorem to find diagonals within P_1 and P_2, and so forth, until the polygon was completely triangulated. Indeed, every algorithm following an inductive construction on n can be viewed as a type of *unbalanced* recursive divide-and-conquer algorithm, with problems of size 1 and $n-1$. Similarly, a divide-and-conquer method can be viewed as induction, although the common practice is to use the term "divide-and-conquer" when the subproblems are of size more than $O(1)$ each.

The divide-and-conquer paradigm partitions the problem into two parts, solves each of the parts recursively, and then "merges" the two solutions to obtain the full solution. In 1977, Franco Preparata and Se June Hong were the first to apply this technique to the convex hull problem. Their goal was to construct a fast algorithm for 3D point sets; although we consider the planar case, there is a natural extension to 3D that is the focus of the next section. Figure 2.10 gives an example of their algorithm in action.

Let S be a point set in general position, with no three points collinear and no two points on the same vertical line. The divide-and-conquer algorithm begins like the incremental algorithm, by sorting the points according to x-coordinate. Divide the points into two (nearly) equal groups, A and B, where A contains the left $\lceil n/2 \rceil$ points and B the right $\lfloor n/2 \rfloor$ points. We then compute the convex hull of A and B recursively (by using the divide-and-conquer algorithm). Finally, we merge conv(A) and conv(B) to obtain conv(S).

Divide and conquer recursively calls itself, with smaller and smaller point sets, until three or fewer points are in each subset. The hull is then immediate. So the recursive step is quite straightforward; the cleverness and geometry come into play in the merge step. Our problem then is to find two tangent lines between the polygons conv(A) and conv(B), one supporting the two convex hulls from below and the other from above. Figure 2.11(a) shows an example of the two tangent lines, with part (b) the merged new polygon. This problem is quite similar to the incremental algorithm, where we needed to find two tangent lines between a *point* and a polygon, as given in Figure 2.4(b). Recall that we found the two tangent lines within $O(n)$ time complexity by walking around conv(H_k). The current challenge is to find tangent lines between *two polygons*, not just a polygon and a point. Thus our time complexity increases to $O(n^2)$,

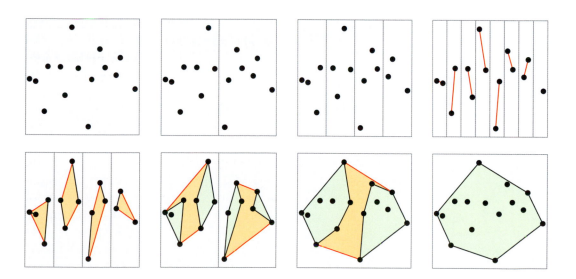

Figure 2.10. The divide-and-conquer algorithm in action.

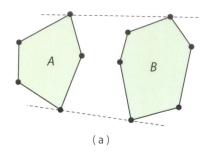   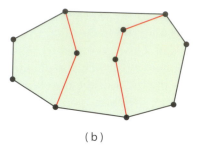

(a)   (b)

Figure 2.11. (a) Finding two tangents and (b) constructing the convex hull.

where for each point of the first polygon, we need to walk around the second one checking for tangent lines.

Exercise 2.31. *It might seem that the highest and lowest points of A and B should always be the points of tangency we are looking for. Find examples where this is not the case.*

Exercise 2.32. *Can you find conditions under which the highest and lowest points of A and B will be always be the tangent points we seek?*

With cleverness and geometric intuition, Preparata and Hong were able to find the tangent lines in linear time. Let's describe how to find the lower tangent line, the one supporting the two polygons from below. The sorting step along with our general position assumption (no two points lie on the same vertical line) guarantees that A is to the left of B, separated by a vertical line. Let α be the rightmost point of A and β the leftmost point of B. Assuming that α is a fixed point (the role of p in the incremental algorithm), proceed by walking counterclockwise from β along the vertices of B. Continue this until a lower tangent line at a vertex of B is found passing through α. Let β be this new vertex of B. Fixing β now, walk clockwise from α around A until a new α is found, which will be a lower tangent to A passing through β. As we repeat this process of walking along A and B, we will eventually reach a lower tangent line supporting both A and B. Figure 2.12 shows an example of this walk.

Exercise 2.33. *Prove that walking along A and B as described above guarantees the lower tangent line being found.*

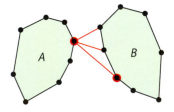   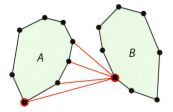   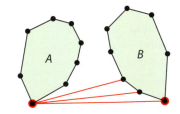

Figure 2.12. Alternately walking between A and B, looking for the lower tangent.

Exercise 2.34. *Show that throughout the walk, the line segment $\alpha\beta$ never intersects the interior of A or B.*

The algorithm for finding the upper tangent line is analogous. The cost of finding the tangent lines is linear, walking around A and B at the same time. Let's now consider the time complexity of the entire divide-and-conquer algorithm. Since this algorithm is recursive, calculating the speed is not straightforward. Let $T(n)$ be the time complexity of the divide-and-conquer hull algorithm for n points. Then $T(n) = 2T(n/2) + O(n)$, where $2T(n/2)$ are the recursion halves and $O(n)$ is the merge step. This is a classical recurrence relation in computer science, whose solution is $T(n) = O(n \log n)$. (The $\log n$ term derives from the fact that n can only be divided in half $\log_2 n$ times before it is reduced to 1 or below.) Thus we have found another convex hull algorithm with optimal time complexity. We summarize it as follows:

DIVIDE-AND-CONQUER   Convex Hull Algorithm $O(n \log n)$

Sort the points of S by x-coordinate. Divide the points into two (nearly) equal groups. Compute the convex hull of each group (recursively using divide and conquer). Merge the two groups together with upper and lower supporting tangents to get the hull of S.

Exercise 2.35. *Analyze the time complexity for triangulating a polygon following the recursive method implied by Theorem 1.4.*

Exercise 2.36. *If the sorting step of the divide-and-conquer algorithm is skipped, the two hulls A and B that result will, in general, intersect. Construct a merge algorithm that can combine two possibly intersecting convex hulls with n and m vertices in $O(n + m)$ time.*

2.7 CONVEX HULL IN 3D

The notion of convexity is fundamentally dimension independent. In particular, it extends to three dimensions and is the basis of the convex hull of a set of points in $\mathbb{R}^3$. This hull is a *convex polyhedra* in 3D, the analog of a convex polygon in 2D. Although we will not explore convex polyhedra in detail until Chapter 6, here we will quickly contrast the computation of the 3D hull with that of the 2D hull.

The convex hull of points in $\mathbb{R}^3$ is a fundamentally more complex object than a convex polygon, composed of vertices, edges, and *faces*, where each face is itself a convex polygon. Figure 2.13 shows the convex hull of 758 random points on the surface of a sphere. Remarkably,

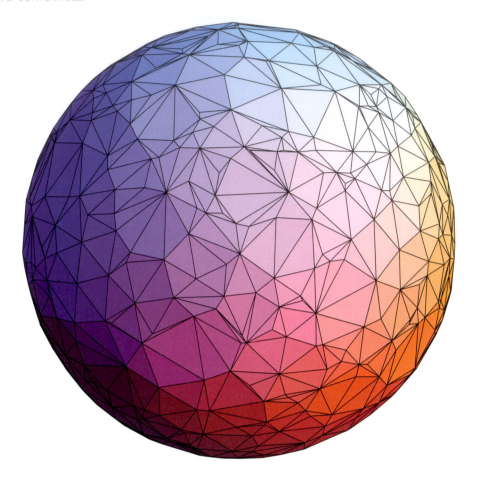

Figure 2.13. The convex hull of 758 random points on the surface of a sphere.

this increased conceptual complexity does not translate into increased computational time complexity.

One might first wonder about the *combinatorial complexity* of the surface of the hull of n points. In 2D, the hull is a polygon of at most n vertices and n edges, and so it is feasible to hope for an algorithm that is close to linear-time complexity. Indeed, we saw two such algorithms, Graham scan and divide-and-conquer, that achieved $O(n \log n)$ time complexity. In 3D, the hull again has at most n vertices, but it has more edges and faces. The example in Figure 2.13 has 758 vertices but 2268 edges and 1512 faces. For n vertices, it turns out that the number of edges is always less than $3n$, and the number of faces is always less than $2n$. These bounds follow from Euler's famous formula, which we will describe and prove in Section 3.1 and again in Section 6.2.

Accepting that the total combinatorial complexity of the surface is of order $O(n)$, independent of which data structure is used to represent the surface, it is still feasible to hope for an algorithm close to linear-time complexity. This reasoning, incidentally, fails in higher dimensions: in 4D, the convex hull can have quadratic complexity. In general, the hull of n points in d dimensions may have $\Omega(n^{\lfloor d/2 \rfloor})$ complexity. The floor function in the exponent saves us for $d = 3$.

We explored four algorithms for constructing the 2D hull. All but one of them extends to 3D, with complexities given in the table below. There

Algorithm	2D Complexity	3D Complexity
Incremental	$O(n^2)$	$O(n^2)$
Gift wrapping	$O(nh)$	$O(nf)$
Divide-and-conquer	$O(n \log n)$	$O(n \log n)$
Graham scan	$O(n \log n)$	?

is no counterpart (or at least none has been discovered) for the Graham scan in 3D, for there appears to be no obvious analog of "clockwise order." Gift wrapping again is output-sensitive: nf depends on the number of faces f in a manner similar to the 2D gift-wrapping algorithm's dependence on the number of hull edges h. We will concentrate on the two most important algorithms, the incremental and the divide-and-conquer.

> **UNSOLVED PROBLEM 10**   **3D Graham Scan**
>
> Find a natural counterpart for the Graham scan algorithm in 3D.

The incremental algorithm is again quadratic, but this time it is often the algorithm of choice due to its conceptual simplicity. Indeed, the hulls in the figures in this section were all computed via an implementation of the incremental algorithm. The overall structure of the 3D incremental algorithm is identical to that of the 2D version: Let Q be the current hull. At each iteration, add one new point p and compute the hull of $Q \cup p$. This becomes the new hull Q' and the process is repeated.

In 2D, this hull computation amounted to finding two tangents from p to Q, as shown in Figure 2.4. In 3D, however, we need to find tangent *planes* rather than tangent lines. These planes bound a *cone* of triangle faces, each of whose apex is p and whose base is an edge e of Q. Figure 2.14(a) shows the hull Q of 100 points in 3D, and part (b) displays the hull of $Q \cup p$. Notice the cone of triangular faces incident to p, where the bases of these triangles are edges of Q.

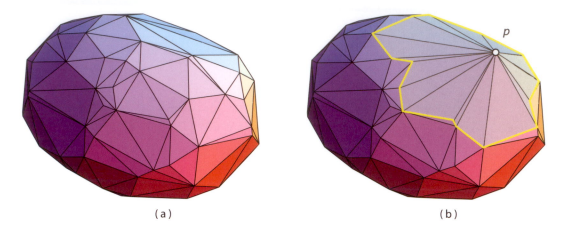

(a)   (b)

Figure 2.14. (a) The hull Q of 100 points and (b) the hull of $Q \cup p$ along with the shadow boundary marked.

Imagine standing at p and looking toward Q. Assume for the moment that no faces are viewed edge-on; in other words, the interior of each face of Q is either visible or not visible from p. It should be clear that the visible faces are precisely those that are to be discarded in moving from Q to Q'. Moreover, the edges on the border of the visible region are precisely those that become the bases of cone faces apexed at p. For suppose e is an edge of Q such that the plane determined by e and p is tangent to Q. Edge e is adjacent to two faces, one of which is visible from p and one of which is not. Therefore e is on the border of the visible region. An equivalent way to view this is to think of a light source placed at p. Then the visible region is that portion of Q illuminated and the border edges are those between the light and dark regions, a type of *shadow boundary*. Again see Figure 2.14(b), which shows the shadow boundary of the point p, and compare it to the 2D case given in Figure 2.4.

From this discussion, it is evident that if we can determine which faces of Q are visible from p and which are not, then we will know enough to find the border edges. This will allow us to construct the cone and discard the appropriate faces of Q. Define a face f to be *visible* from p if some point x interior to f is visible from p, that is, px does not intersect Q except at x. Note that under this definition, seeing only an edge of a face does not render the face visible, and faces seen edge-on are also considered invisible. Just as in 2D, this visibility calculation is an $O(1)$ constant-time computation.

Exercise 2.37. *Detail a method to determine whether a triangle face $f = (a, b, c)$ is visible from p.*

This then gives us an algorithm: For each face, compute if it is visible or not. Those edges e adjacent to both a visible and an invisible face lead to a cone face apexed at p. Then all the visible faces can be discarded, and we have the new hull Q'. The addition of each point p can be accomplished in $O(n)$ time, so the algorithm achieves $O(n^2)$ time overall.

Exercise 2.38. *Let Q be a regular tetrahedron and p a point outside Q. What is the greatest number of faces conv($Q \cup p$) can have for any p? What is the fewest? Can conv($Q \cup p$) have an odd number of faces?*

Exercise 2.39. *Prove that the boundary edges of the region of Q which are visible to p form a simple (nonintersecting) closed curve.*

★ **Exercise 2.40.** *Provide a version of the gift-wrapping algorithm to compute the convex hull in 3D.*

The divide-and-conquer paradigm is the same as in two dimensions: sort the points by their x-coordinate, divide into two sets, recursively construct the hull of each half, and merge. The merge must be accomplished in $O(n)$ time to achieve the desired $O(n \log n)$ bound. All the work is in the merge, and we concentrate solely on this.

Let A and B be the two hulls to be merged. The hull of $A \cup B$ adds a "band" of faces, as shown in Figure 2.15. The number of these faces will be linear in the size of the two polyhedra: each face uses at least one edge of either A or B, so the number of faces is no more than the total number of edges. So it is feasible to perform the merge in linear time, as long as each face can be added (on average) in constant time.

Let π be a plane that supports A and B from below, touching A at a vertex p and B at a vertex q. (The 2D version of this is given in Figure 2.11, where the supporting plane π corresponds to a tangent line.) To make the exposition simpler, assume that p and q are the only points of contact of π. Then π contains the line L determined by pq. Now "crease the plane" along L and rotate half of it about L until it bumps into one of the two polyhedra.

A crucial observation is that if it first bumps into a point r on a polyhedron, say A, then pr must be an edge of A. In other words, the first point r hit by π must be a neighbor of either p or q. This limits the vertices that need to be examined to determine the next to be bumped. Once π hits r, one triangular face of the merging band has been found: the triangle pqr. Now the procedure is repeated, but this time around the line through qr (since r is on A). The wrapping stops when it closes upon itself.

After wrapping around A and B with a cylinder of faces, it only remains to discard the faces hidden by the wrapped band to complete the merge. Unfortunately, the wrapping process does not immediately tell us which faces of A are visible from some point of B, and vice versa; it is

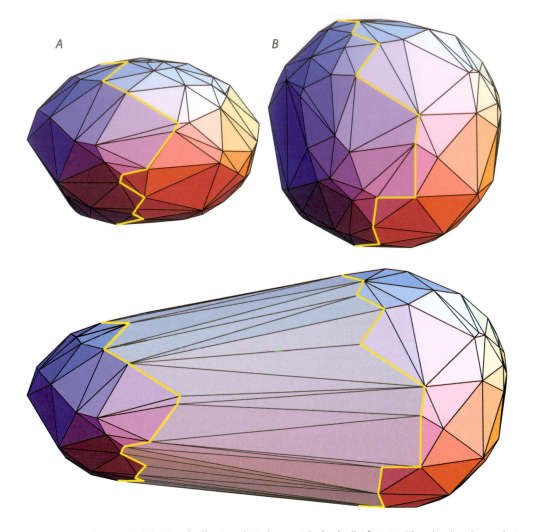

Figure 2.15. Two hulls A and B along with the hull of $A \cup B$. The shadow boundaries are marked.

just these faces that should be deleted. But the wrapping does discover all the "shadow boundary" edges: those edges of A and B touched by one of the wrapped faces. (If all of B were a light source, the shadow boundary on A marks the division between light and dark; and symmetrically the shadow boundary on B separates light from dark when A is luminous.) Intuitively one could imagine "snipping" along these edges and detaching the hidden caps of A and B.

Alas, contrary to intuition, the shadow boundary edges on polyhedra A and B do not necessarily form simple cycles! So even this seemingly straightforward step is tricky to implement. Nevertheless, a careful analysis shows that the merge is $O(n)$, which leads to $O(n \log n)$ time. Despite the asymptotic advantage of this algorithm over the incremental algorithm, the delicacy of implementing the wrapping and updating the

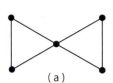 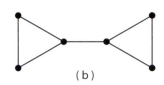 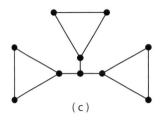

Figure 2.16. Possible shadow boundary edges of polyhedra.

surface data structure has left this algorithm theoretically important but not the pragmatic choice.

Exercise 2.41. *Prove that the faces deleted from A during the merge step form a connected set.*

Exercise 2.42. *For each shape shown in Figure 2.16, construct examples of convex polyhedra A and B such that the shadow boundary edges of A in* conv$(A \cup B)$ *has that particular shape. This shows the complexity of 3D divide-and-conquer as compared to 3D incremental, as contrasted with Exercise 2.39.*

Exercise 2.43. *Let A and B be congruent nonintersecting cubes, with B a translated copy of A. What is the most faces* conv$(A \cup B)$ *can have? What is the fewest?*

Exercise 2.44. *Let A and B be congruent nonintersecting regular icosahedra, with B a translated copy of A. Arrange A in some convenient position (say with the line through the top and bottom vertices vertical), and choose some convenient direction for the translation (say perpendicular to some edge and to the vertical). Describe* conv$(A \cup B)$.

SUGGESTED READINGS

Alexander Barvinok. *A Course in Convexity*. American Mathematical Society, 2002.
This well-written textbook covers the applications of convexity to numerous areas of mathematics and computer science, such as sphere packing, graph theory, linear programming, and polyhedra. It is aimed at an advanced undergraduate level.

Joseph O'Rourke. *Computational Geometry in C*. Cambridge University Press, 2nd edition, 1998.
Chapters 3 and 4 of this text cover convex hulls in 2D and 3D, respectively, with an emphasis on implementations.

Franco Preparata and Michael Shamos. *Computational Geometry: An Introduction*. Springer-Verlag, 3rd edition, 1990.
The first textbook in computational geometry (initially published in 1985). Based on Shamos's seminal Ph.D. thesis from 1978, it contains a detailed description of the Preparata and Hong 3D hull algorithm.

Stefan Schirra. Robustness and Precision Issues in Geometric Computation. In Jörg-Rüdiger Sack and Jorge Urrutia, editors, *Handbook of Computational Geometry*, chapter 14, pages 597–632. Elsevier, 2000.
A survey of work confronting degeneracies and other low-level computational issues, exploring the two main options of exact geometric computation.

Lutz Kettner and Stefan Näher. Two Computational Geometry Libraries: LEDA and CGAL. In J. E. Goodman and J. O'Rourke, editors, *Handbook of Discrete and Computational Geometry*, chapter 65, pages 1435–1463. CRC Press LLC, 2nd edition, 2004.
Both the LEDA and CGAL libraries offer options for robust computation.

Raimund Seidel. Convex hull computations. In J. E. Goodman and J. O'Rourke, editors, *Handbook of Discrete and Computational Geometry*, chapter 22, pages 495–512. CRC Press LLC, 2nd edition, 2004.
A survey of convex hull algorithms written by one of the originators of these algorithms. Especially good coverage of algorithms for higher dimensions.

TRIANGULATIONS $\mathbf{3}$

The previous chapter concentrated on finding the boundary of a point set, represented by its convex hull. In this chapter, we focus on the interior of a point set, partitioning it into triangles in a manner similar to the triangulation of a polygon in Chapter 1. But triangulating a structureless point set differs in many regards from triangulating a polygon.

We start with basic algorithms and combinatorics (Section 3.1). Then a higher-level view shows that the space of all triangulations of a fixed point set has a rich and beautiful structure encapsulated in the flip graph (Section 3.2). Here we take a detour into a non-traditional but fascinating topic, the associahedron (Section 3.3). This uncovers an even deeper structure to the triangulations of a set of points in convex position, showing that the flip graph has the structure of a polyhedron. We then concentrate on what is arguably the most important triangulation, the Delaunay triangulation (Section 3.4), which has striking properties and plays a central role in many application areas. Finally, we touch on a miscellany of special triangulations, ending with the most recently discovered, pseudotriangulations (Section 3.5).

3.1 BASIC CONSTRUCTIONS

When discussing polygons, we distinguished between boundary edges and internal diagonals. For a point set S, the term *edge* is used to indicate any segment that includes precisely two points of S at its endpoints.

Definition. A *triangulation* of a planar point set S is a subdivision of the plane determined by a maximal set of noncrossing edges whose vertex set is S.

The word *maximal* in the definition indicates that any edge not in the triangulation must intersect the interior of at least one of the edges in the triangulation. Figure 3.1(a) shows the convex hull of a point set; (b) shows a subdivision of the hull into triangles, where the three marked points are collinear. This is not a maximal subdivision, however, because we can draw another edge that avoids the preexisting edges, as in part (c). Parts (d) and (e) show two different triangulations.

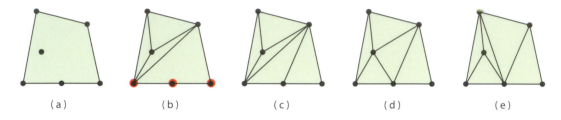

(a)   (b)   (c)   (d)   (e)

Figure 3.1. A point set S along with (a) its convex hull, (b) a subdivision, and (c)–(e) three different triangulations of S.

Exercise 3.1. *Find all the distinct triangulations of the point set in Figure 3.1.*

Exercise 3.2. *Show that the edges of the convex hull of a point set S will be in every triangulation of S.*

Exercise 3.3. *The definition of a triangulation of a point set does not even mention "triangles." Show that all the regions of the subdivision inside the convex hull must indeed be triangles.*

We now discuss a simple algorithm for constructing triangulations which we call the *triangle-splitting* algorithm. Assume for simplicity that our points are in general position, with no three points collinear. Begin by finding the convex hull of the point set and triangulate this hull as a polygon (ignoring the interior points). Note that each interior point is strictly inside some triangle (and not on a diagonal) due to the general position assumption. Choose an interior point and draw edges to the three vertices of the triangle that contains it. Repeating this process until all interior points are exhausted produces a triangulation.

Figure 3.2 shows three rows of examples of this procedure. The first column shows the triangulation of the convex hull polygon and the last column the final triangulation following the algorithm. Notice that both the initial hull triangulation and the order of processing the interior point affects the final triangulation obtained.

TRIANGLE-SPLITTING   Triangulation Algorithm

Find the convex hull of S and triangulate this hull as a polygon. Choose an interior point and draw edges to the three vertices of the triangle that contains it. Continue this process until all interior points are exhausted.

Exercise 3.4. *Extend this algorithm to work for points that may include three or more collinear points (but not all collinear).*

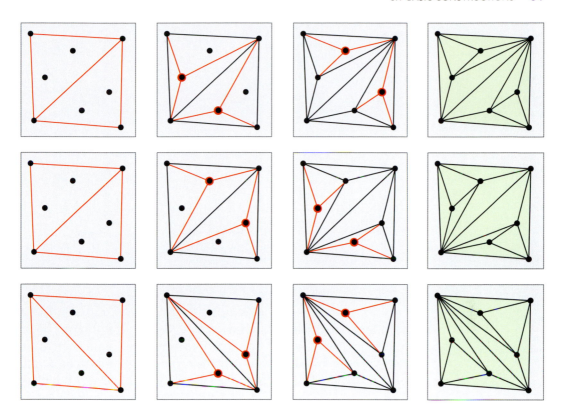

Figure 3.2. The rows show the same algorithm producing three different tri-angulations.

Exercise 3.5. *Analyze the time complexity of the triangle-splitting algorithm.*

We have seen that a point set can have many distinct triangulations. But looking at the last column of Figure 3.2, each triangulation has exactly ten triangles. Indeed, for triangulations of a point set created from the triangle-splitting algorithm, a natural count of the number of triangles immediately follows from the algorithm.

Lemma 3.6. *Let S be a point set of k points in the interior and h points on the hull. If not all points are collinear, any triangulation of S that results from the triangle-splitting algorithm has exactly $2k + h - 2$ triangles.*

Proof. By Theorem 1.8, the triangulation of the convex hull of S has $h - 2$ triangles. If an interior point is within a triangle, the algorithm connects this point to the three vertices of the triangle containing it. Thus, such an interior point replaces one triangle by three triangles, increasing the triangle count by +2. But if an interior point lies on an edge, the algorithm (extended by Exercise 3.4) connects this point to the two

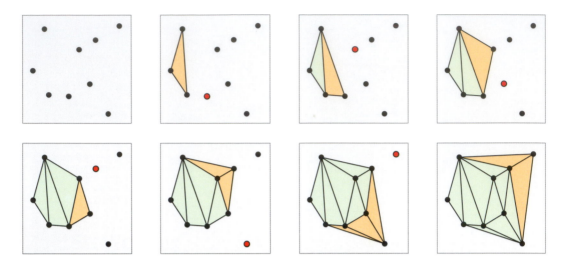

Figure 3.3. The incremental triangulation algorithm in action.

vertices of the triangles on either side of this edge. So this interior point splits the triangles on either side into two triangles, again increasing the triangle count by $+2$. Since there are k interior points in S, the number of triangles resulting from the triangle-splitting algorithm must be $2k + h - 2$.   □

Exercise 3.7. *Prove or disprove: The triangle-splitting algorithm produces all possible triangulations of a point set.*

A second algorithm may be obtained by extending the incremental algorithm we explored for convex hulls (Section 2.2) to triangulations. The extension is almost immediate. Figure 3.3 shows this algorithm in action. At each step, the next point p is connected to the previous triangulated convex polygon by a "fan" of triangles apexed at p.

INCREMENTAL   Triangulation Algorithm

Sort the points of S according to x-coordinates. The first three points determine a triangle. Consider the next point p in the ordered set and connect it with all previously considered points $\{p_1, \ldots, p_k\}$ which are *visible* to p. Continue this process of adding one point of S at a time until all of S has been processed.

Exercise 3.8. *Analyze the time complexity of the incremental algorithm.*

Exercise 3.9. *Prove that the incremental algorithm produces a triangulation of a point set in general position.*

Exercise 3.10. *Prove or disprove: The incremental algorithm produces all possible triangulations of a point set S, assuming all possible rotations of S.*

Exercise 3.11. *Alter the Graham scan convex hull algorithm (Section 2.4) to compute triangulations.*

The number of triangles in any triangulation of a polygon depends only on the number of vertices of the polygon. What is the situation for triangulations of a point set? Lemma 3.6 shows that any triangulation of a point set deriving from the triangle-splitting algorithm has a fixed number of triangles, dependent on the number of interior points k and the number of hull points h of the set. We show below that this same result holds true for *every* triangulation of a point set. The proof uses a beautiful and powerful formula of Leonhard Euler, the brilliant eighteenth-century Swiss mathematician. This formula will be examined in detail (and reproved) when we discuss polyhedra in Chapter 6, but here we offer a concise proof for plane graphs without connecting it to its original polyhedral context.

Theorem 3.12 (Euler's Formula). *Let G be a connected planar graph with V vertices, E edges, and F faces on the plane (where the outer face is unbounded). Then $V - E + F = 2$.*

Proof. We prove this using induction on the number of edges. If $E = 0$, then G is an isolated vertex on the plane and $V - E + F = 1 - 0 + 1 = 2$. Otherwise, choose any edge e of G. If e connects two vertices of G, contract this edge, reducing V and E by one. If e is incident to only one vertex (i.e., e is a *loop* separating two faces), delete this edge, reducing F and E by one. In either case, the new graph remains connected and the formula follows by induction. $\square$

Notice that the Jordan curve theorem of Chapter 1 is used in this proof when we assume any loop separates two faces of G. Figure 3.4 shows a progression of examples based on the proof above. Contraction of the red edge or deletion of the red loop does not change the value $V - E + F$ of the graph.

We now apply Euler's formula to triangulations of point sets. The following theorem not only proves that each triangulation must have the same number of triangles, but provides an explicit count of the exact

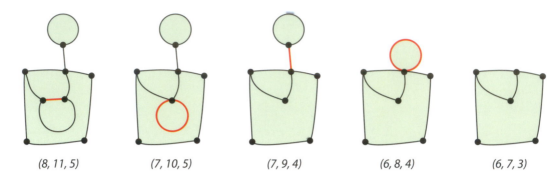

| (8, 11, 5) | (7, 10, 5) | (7, 9, 4) | (6, 8, 4) | (6, 7, 3) |

Figure 3.4. Contraction and deletion of edges of a connected planar graph. The respective values (V, E, F) are given below each graph.

number of triangles. Euler's formula is in some sense a generalization of Theorem 3.13 to arbitrary subdivisions.

Theorem 3.13. *Let S be a point set, with h points on the hull and k in the interior, and so $n = k + h$ total. If not all points are collinear, then any triangulation of S has exactly $2k + h - 2$ triangles and $3k + 2h - 3$ edges.*

Proof. Let T be a triangulation of the point set S, and let t be the number of triangles of T. We know T subdivides the plane into $t + 1$ faces, t triangles inside the hull and the face outside the hull. Each triangle has three edges, and the outside face has h edges. Since each edge touches exactly two faces, $(3t + h)$ double counts edges; so there are exactly $E = (3t + h)/2$ edges. Applying Euler's formula with $V = n$ and $F = t + 1$ results in

$$n - \frac{1}{2}(3t + h) + (t + 1) = 2.$$

Solving for t yields

$$t = 2n - h - 2 = 2k + h - 2.$$

We can then use this value for t to find the number of edges in terms of k and h. □

Exercise 3.14. *Show that every triangulation has some vertex of degree at most five.*

Having settled that each triangulation of a fixed point set S has the same number of triangles, it is now natural to turn to counting the number of distinct triangulations of S, a problem we investigated for triangulations of polygons in Section 1.2. This counting question seems difficult to answer, for it depends on the distribution of the points of S with respect to one another. For example, we know from Theorem 1.19

that if S is a point set in general position with all points on the hull (the points are in *convex position*, to use the technical term), then the number of triangulations of S is the Catalan number. Only such special cases of point sets are fully understood.

Exercise 3.15. *Construct a point set S with $n + 2$ points such that the number of triangulations of S is greater than the Catalan number C_n.*

Even obtaining a good upper bound seems difficult. The following result of Micha Sharir, Adam Sheffer, and Emo Welzl from 2009, established in a 60-page paper, is the best available, but it is not known to be tight.

Theorem 3.16. *Let S be a planar point set of n points. Then S has no more than 30^n distinct triangulations.*

Exercise 3.17. *Compare this value with the Catalan number derived in Theorem 1.19.*

Exercise 3.18. *Consider the point set S given in Figure 3.5. It is made of two "double chains" of points, where every pair of points from different chains is visible to each other. Show that the edges drawn in the figure appear in every triangulation of S. Moreover, if there are n points in each chain, find the number of triangulations of S.*

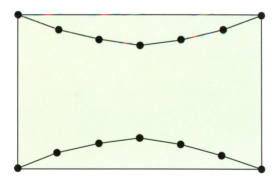

Figure 3.5. A double-chain point set.

UNSOLVED PROBLEM 11   **Triangulation Algorithm**

Find an algorithm for counting the number of triangulations of an n-point set in the plane that runs in time polynomial in n. Because there are an exponential number of triangles, a solution must count without effectively listing them all.

3.2 THE FLIP GRAPH

One method of exploring the structure of a point set S is to examine triangulations of S. We just looked at counting such triangulations, but this is a crude probe. The set of triangulations has a rich structure if viewed the right way. Consider the three triangulations shown earlier in Figure 3.1 (c)–(e). Notice that parts (c) and (d) differ by just one diagonal; thus they are closely related as triangulations (like siblings). Similarly, parts (d) and (e) also differ by one diagonal. However, (c) and (e) differ by two diagonals, making them cousins. This relationship between triangulations of a point set can be made precise; we first need a definition.

Consider the triangulation of the convex quadrilateral $ABCD$ shown in Figure 3.6(a). An *edge flip* (called a *flip* for short) removes the diagonal AC and replaces it with diagonal BD, as illustrated in (b). Flipping BD again brings us back to (a). Flips are not possible for nonconvex quadrilaterals as shown in (c). Flips permit defining a relationship between two triangulations of a point set.

Definition. For a point set S, the *flip graph* of S is a graph whose nodes are the set of triangulations of S. Two nodes T_1 and T_2 of the flip graph are connected by an arc if one diagonal of T_1 can be flipped to obtain T_2.

Figure 3.7 shows the flip graph of a point set of six points. Each red node corresponds to the triangulation drawn within the node. The flip graph has nine nodes and eleven arcs.

Exercise 3.19. *For every n, construct a point set (not necessarily in general position) with n points whose flip graph is a single node, that is, no flip is possible.*

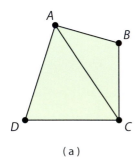

(a)

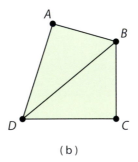

(b)

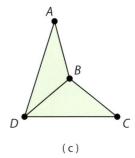

(c)

Figure 3.6. An edge flip is possible for convex quadrilaterals, but not for nonconvex ones.

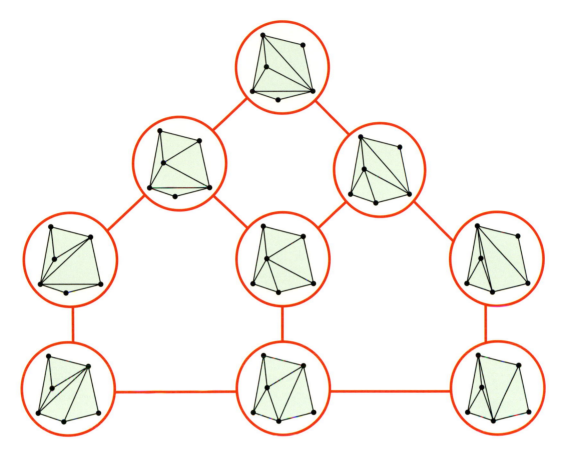

Figure 3.7. An example of the flip graph of a point set.

Exercise 3.20. *For every n, construct a point set (not necessarily in general position) with n points whose flip graph is two nodes connected by an arc.*

Exercise 3.21. *Figure 3.8 shows a triangulation for three different point sets. For each point set, find its flip graph.*

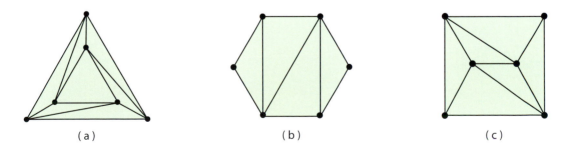

    (a)   (b)   (c)

Figure 3.8. Triangulations of three point sets.

The flip graph reveals a discrete *space of triangulations* of a fixed point set. There are several natural questions about triangulations that can be interpreted in the language of flip graphs. The most basic of these is whether one triangulation of S can be transformed into another via a sequence of flips. In other words, is the flip graph connected? The following theorem, first proved by Charles Lawson in 1971, answers YES.

Theorem 3.22. *The flip graph of any point set in the plane is connected.*

Proof. Let S be a planar point set with n points. Order the points of S according to x-coordinates. If two or more points share the same x-coordinate, we can always rotate the plane slightly such that all points have distinct x-coordinates. Label the resulting ordering of S as $\{p_1, \ldots, p_n\}$. Let T_* be the triangulation obtained from S using the incremental algorithm for triangulations described in Section 3.1 above. Our goal is to show that any triangulation T of S can be converted into T_* using flips. This will prove the connectivity of the flip graph because any two triangulations will be connected by paths to the T_* node of the flip graph, and thus to each other by reversing the flips in one of the paths.

We prove this by induction on n. When $n = 3$, the set S has a unique triangulation, its flip graph is a single node, and we are done. Assume for any point set S with fewer than n points, any triangulation of S can be made into the incremental algorithm triangulation of S by flips. Now consider S with n ordered points $\{p_1, \ldots, p_n\}$ and let T be a triangulation of S. Let the *star* of a vertex v of a triangulation be the union of the triangles incident to v. We will show that by a sequence of flips, the star of p_n in our triangulation T (shown at the left in Figure 3.9) can be converted into the star of p_n in T_* (shown at the right of the figure). Once this is accomplished, what remains is a triangulation of the point set $S \setminus \{p_n\}$, which by our induction hypothesis

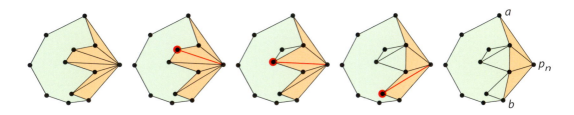

Figure 3.9. Flipping edges to obtain another triangulation.

Because the incremental algorithm produces a convex polygon at each step of the process (see Figure 2.3), the star of p_n in T_* has exactly three convex vertices: p_n and the two vertices adjacent to p_n on the hull, call them a and b. In T_*, the vertices between a and b in the star form a reflex chain. Now choose a convex vertex v of the star of p_n in our triangulation T distinct from p_n, a, and b. If there is no such convex vertex, then the star of p_n in T is exactly the star of p_n in T_*, and so we are finished. Because v is convex, the edge between v and p_n is a diagonal of a convex quadrilateral that can be flipped. Thus one vertex fewer is now visible to p_n and the degree of p_n decreases by one. Repeat this process choosing such convex vertices. This process must end since the degree of p_n is decreasing at each step.   □

Exercise 3.23. *Prove that if p_n, a, and b are the only convex vertices in the star of p_n, then the star of p_n in T is exactly the star of p_n in T_*.*

Exercise 3.24. *Prove or disprove: no point set can have a triangle as a subgraph of its flip graph, that is, three nodes connected in a cycle by three arcs.*

Not only is the connectivity of the flip graph an elegant geometric result, it has practical implications as well. It shows that one triangulation can be converted to another by local moves, one step at a time. Indeed, it forms the basis of many algorithms that incrementally "improve" a triangulation, as we will see in Section 3.4.

Now that we know the flip graph is connected, it is natural to wonder about its shape. A quantity that give some hint in this direction is the diameter. The *diameter* of a graph is the longest path between any two nodes of the graph, where the length of the path is its number of arcs. For a complete graph, where there is an edge between any two nodes, the diameter is one. So the diameter gives a sense of how densely a graph is connected. The proof of the previous theorem provides us with an upper bound on the flip graph's diameter.

Corollary 3.25. *For a planar point set S of n points, the diameter of its flip graph is at most $(n-2)(n-3)$.*

Proof. We show by induction that any triangulation can be converted into T_* (the incremental triangulation) with at most $\binom{n-2}{2}$ flips. Consider the proof of Theorem 3.22. To reach the star of p_n in T_*, at most $n-3$ flips were needed. By the induction hypothesis, the remaining number of flips needed will at most be $\binom{n-3}{2}$. An explicit calculation shows $\binom{n-3}{2} + (n-3) = \binom{n-2}{2}$. So any two nodes of the flip graph can therefore be connected by $\binom{n-2}{2}$ arcs to T_*, and so by twice this number, which is $(n-2)(n-3)$, to each other. □

Exercise 3.26. *Find the diameters of the flip graphs of the point sets given in Figure 3.8.*

★ **Exercise 3.27.** *Find the diameter of the flip graph of the point set given in Figure 3.5.*

Given two triangulations T_1 and T_2 of a point set with n points, what is the least number of flips needed to convert one into the other? In other words, what is the length of the shortest path from T_1 to T_2 in the flip graph? The diameter yields an upper bound of $(n-2)(n-3)$, but a novel bound was obtained by Sabine Hanke, Thomas Ottmann, and Sven Schuierer in 1996:

Theorem 3.28. *Let S be a point set in general position, and let T_1 and T_2 be two triangulations of S. Let T_{12} be the diagram obtained by overlapping the triangulations T_1 and T_2. Then the distance between T_1 and T_2 in the flip graph is at most the number of crossings between edges in T_{12}.*

Exercise 3.29. *Figure 3.10 shows two triangulations from Figure 3.2 along with their overlapping diagram with 16 edge crossings. Verify Theorem 3.28 by finding a path in the flip graph between the two triangulations with no more than 16 flips.*

★ **Exercise 3.30.** *Let S be a point set with h points on the hull and k points interior to the hull. Prove that the number of crossings of edges in T_{12} is at most $(3k + h - 3)^2$.*

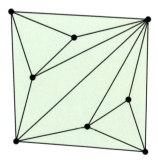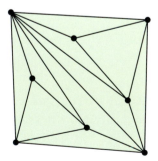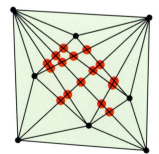

Figure 3.10. Triangulations T_1 and T_2 along with the overlapping diagram T_{12}.

> **UNSOLVED PROBLEM 12**   Shortest Path
>
> Find a polynomial-time algorithm that finds a shortest path between any two nodes of the flip graph of S.

Now that we have considered triangulations of point sets in the plane, let's think about how the situation changes in three dimensions. Given a point set S in $\mathbb{R}^3$, we seek tetrahedralizations of S. Our experience with untetrahedralizable polyhedra in Chapter 1 (see Figures 1.7 and 1.17) might suggest that not every point set can be tetrahedralized. But in fact every point set can: a variation on the incremental algorithm for the 3D convex hull achieves tetrahedralization in a manner similar to the 2D incremental triangulation algorithm in Section 3.1. Sort the points using *lexicographical ordering*: sort by x, and among those points with the same x-coordinate, sort by y, and among those with the same x- and y-coordinates, sort by z. As each point p is added one at a time, edges are drawn from p to the vertices of the convex hull of the previous points that are visible to p.

Exercise 3.31. Show that this incremental algorithm indeed results in a partition of the convex hull into tetrahedral pieces.

Whereas Theorem 3.13 provides an exact count of triangles for planar point sets, the number of tetrahedra in different tetrahedralizations of the same point set can vary. This should not come as a surprise as we have already seen that the number of tetrahedra in a polyhedron can differ (Figure 1.11). However, we can obtain some quantitative information about tetrahedra for point sets, as the following theorem shows. We restrict ourselves to points in general position, which in this context means that no three points are collinear and no four coplanar.

Theorem 3.32. Let S be a point set in $\mathbb{R}^3$ in general position, with k points in the interior and h on the hull. Then there exists a tetrahedralization of S with at most $3k + 2h - 7$ tetrahedra.

Proof. Consider the hull of S. Let e be the number of edges and t be the number of triangles in the hull. Note that all faces of the hull are triangles by the general-position assumption. Because each edge borders two triangles, $2e = 3t$. Moreover, by Euler's formula, $h - e + t = 2$. Thus the number of edges and triangles of the hull must be $e = 3h - 6$ and $t = 2h - 4$.

Let v be a vertex of the hull with r incident hull triangles. Construct a tetrahedralization of S by adding an edge from v to every other vertex of the hull of S. (These edges miss interior points by the general-position assumption.) This produces $(2h - 4) - r$ tetrahedra, one for each triangle not incident to v. Choose any interior point and connect it to the four vertices of its containing tetrahedron, subdividing it into four tetrahedra. So this interior point replaces one tetrahedron by four tetrahedra, increasing the tetrahedral count by $+3$. Repeat this process for each interior point in any order. The resulting total number of tetrahedra is

$$(2h - 4 - r) + 3k = 3k + 2h - 4 - r \leq 3k + 2h - 7,$$

where the final inequality follows because $r \geq 3$.   □

Exercise 3.33. *If the point set S were not in general position, how would this affect the number of tetrahedra?*

Exercise 3.34. *Based on the proof of the theorem, how can we find tetrahedralizations with fewer than $3k + 2h - 7$ tetrahedra?*

Exercise 3.35. *For arbitrary $n > 4$, construct an n-point set in $\mathbb{R}^3$ (not necessarily in general position) with a tetrahedralization having at most $n - 3$ tetrahedra.*

Just as in two dimensions, it is possible to structure the space of all tetrahedralizations of a fixed point set with a graph. To move from one node to another, we do not flip an edge, but rather flip a face. The *flip* in this case is illustrated in Figure 1.11 from Chapter 1, where the two tetrahedra forming a 5-vertex convex polyhedron in part (a) of that figure become the three tetrahedra forming this polyhedron in part (b), and vice versa. When points are not in general position (with four or more coplanar), the flips might take two tetrahedra to two other tetrahedra. An example of a flip graph for six points in $\mathbb{R}^3$ (whose convex hull is the triangular prism) is shown in Figure 3.11.

Similarly, one may define flip graphs for point sets in higher dimensions. As in 2D, the most fundamental question is whether or not these flip graphs are connected. The answer to this question for points sets in $\mathbb{R}^d$ was unknown for almost 30 years after the $d = 2$ case was settled (Theorem 3.22). In 2000, Francisco Santos proved that the flip graph is disconnected for point sets when $d \geq 5$. It is still an open question for dimensions three and four. In fact, it is not even known whether the flip graph might contain an isolated node in $\mathbb{R}^3$. The 3D case is especially important because it plays a role in improving the "quality" of the tetrahedra in a meshing of solid objects.

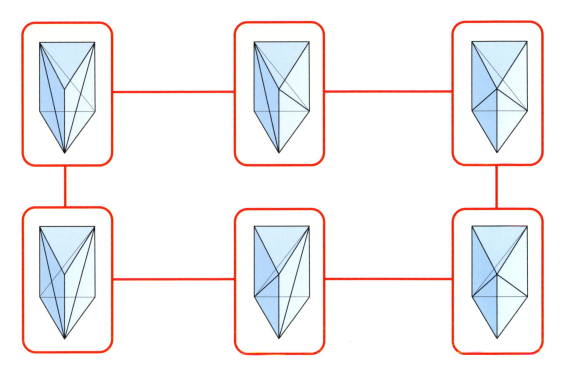

Figure 3.11. An example of the flip graph of six points in $\mathbb{R}^3$.

UNSOLVED PROBLEM 13   Flip Graph in 3D

Is the flip graph of every point set in $\mathbb{R}^3$ connected? If not, is there a set with an isolated node?

Exercise 3.36. *Let T be a point set consisting of the four vertices of a regular tetrahedra. Is it possible to add points P interior to T such that the tetrahedralization of the point set $T \cup P$ results in only regular tetrahedra?*

★ **Exercise 3.37.** *Construct the flip graph of eight points in $\mathbb{R}^3$ whose convex hull forms a cube.*

3.3 THE ASSOCIAHEDRON

For any point set S in the plane, our interests so far have been on the underlying flip graph structure based on triangulations of S. This section focuses on a special configuration of points, points in *convex position*, where the points of S form the vertices of a convex polygon. Here a deeper structure exists which elegantly generalizes the flip graph to higher dimensions. We remark that this section is a digression into advanced material where we catch a glimpse of some ideas on the cutting edge of research. However, we believe the foray is well worth the effort, for we will see triangulations, convex hulls, and polyhedra magically come together.

Exercise 3.38. *Let S be a point set where no three points are collinear. Show that S is in convex position if and only if there is a unique polygon whose vertices are precisely S.*

For the remainder of this section, assume S is a point set in convex position, where no three points are collinear. Now consider the flip graph of S; the nodes of the flip graph correspond to the triangulations of the polygon $P = \text{conv}(S)$. We have already discussed triangulations of a convex polygon in Theorem 1.19, leading to the Catalan number. Thus the flip graph of S will have a Catalan number of nodes.

Let's begin by looking at some examples. When S has three points, P is a triangle, and the flip graph of S is just one node. When S has four points, P is a convex quadrilateral. The flip graph of four points in convex position is a single arc, whose two endpoints correspond to the two ways of triangulating a quadrilateral. The first interesting situation arises for convex polygons with five vertices. Figure 3.12(a) shows the flip graph of a regular pentagon, whose graph is a cycle with five nodes. Upon close examination, it turns out that more structure is hidden beneath the surface than just the flip graph. We first need a definition.

Definition. A *diagonalization* of a polygon P is a decomposition of P into smaller polygons by a set of noncrossing diagonals.

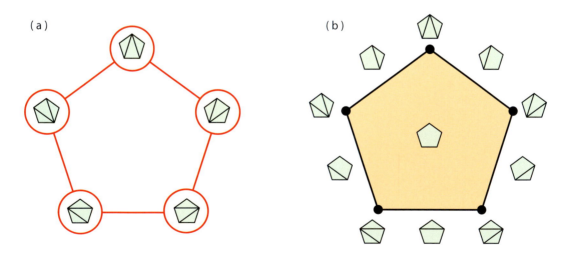

Figure 3.12. (a) The flip graph of a convex pentagon and (b) the 2D associahedron.

Thus a diagonalization can be thought of as a generalization of a triangulation without the "maximal" criterion. Then the polygon is decomposed into smaller polygons, which are not necessarily triangles. Consider the pentagon graph in Figure 3.12(b). We label the nodes with triangulations (diagonalizations using two diagonals), and we can label the edges with diagonalizations using just one diagonal. And the interior of the pentagon graph can be labeled with a diagonalization using no diagonal.

To get a firmer grasp on this structure, consider the case when P is a hexagon. Figure 3.13(a) shows the flip graph of the hexagon embedded in the plane. Part (b) of the figure shows a polyhedron whose *1-skeleton* — its vertices and edges — is exactly this flip graph! Moreover, this polyhedron has nine polygonal faces, each one corresponding to different diagonalizations of the hexagon using one diagonal: The six pentagonal faces are based on the six ways a hexagon is cut into a triangle and a pentagon using one diagonal, whereas the three square faces are based on the six ways a hexagon may be cut into two quadrilaterals using one diagonal. Indeed, there is a rich combinatorial framework embodied by this polyhedron.

These objects we have been looking at are called *associahedra*. A 1D associahedron is the line segment (seen as the flip graph of the quadrilateral). A 2D associahedron is the pentagon in Figure 3.12 and we just described a 3D associahedron in Figure 3.13. What is truly beautiful is that associahedron exists not just for these three examples but higher dimensions as well. The following result was independently proven by Mark Haiman (unpublished) in 1984 and Carl Lee in 1989:

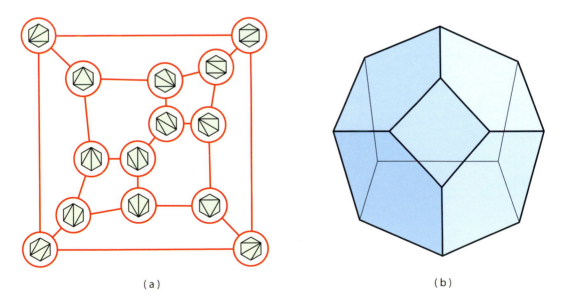

(a)   (b)

Figure 3.13. (a) The flip graph of a convex pentagon and (b) the 3D associahedron.

Theorem 3.39. *There exists a convex n-dimensional polytope[1] called the associahedron whose vertices and edges form the flip graph of a convex (n + 3)-sided polygon. The k-dimensional faces of this polytope are in one-to-one correspondence with the diagonalizations of the polygon using exactly n − k diagonals.*

Notice that the vertices of this polytope (the "faces" of dimension zero) are exactly the triangulations of the polygon, enumerated by the Catalan number. Figure 3.14 provides three different viewpoints of a metal sculpture constructed by Eric Jonash and Sam Kapala that shows the flip graph of a heptagon. This forms the skeletal structure of the 4D associahedron.

Almost twenty years before Theorem 3.39 was discovered, the associahedron had originally been defined by James Stasheff for use in the subfield of topology called homotopy theory. Associahedra have continued to appear in a variety of mathematical fields, currently leading to numerous generalizations. We will see more of these polytopes in Chapter 7, where we discover how *associahedra* are related to *associativity* properties.

Exercise 3.40. *Let P be a polygon with selected diagonals prescribed, as illustrated in Figure 3.15. Draw the flip graph of the polygon with the constraint that these red diagonals are fixed, not flippable.*

★ **Exercise 3.41.** *Let P be a convex polygon with n vertices. Find a formula for the number of diagonalizations of P with exactly k diagonals. Note that when k = n − 3, we obtain the Catalan numbers.*

Figure 3.14. Three perspectives of a metal sculpture of the skeletal structure of the 4D associahedron.

---

[1] A *polytope* is a generalization of the idea of a polygon and a polyhedron to *n* dimensions. We will see more polytopes in Chapter 6.

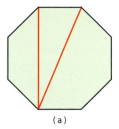 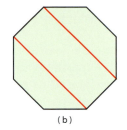 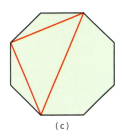

(a)   (b)   (c)

Figure 3.15. Polygons with fixed, unflippable diagonals.

For an n-point set, Corollary 3.25 provides an upper bound of $(n-2)(n-3)$ for the diameter of the flip graph. For the special case of a convex polygon, the diameter of its flip graph is much smaller, and bounded by an impressive result of Daniel Sleator, Robert Tarjan, and William Thurston from 1986, proved via hyperbolic geometry:

Theorem 3.42. *For large enough values of n, the diameter of the flip graph of a convex polygon with n vertices is $2n - 10$.*

From the perspective of associahedra, this result shows the maximum number of steps needed to go from any vertex of the associahedron to any other vertex by walking along the edges. However, this theorem only guarantees diameter "for large values of n." How large n needs to be for the theorem claim to hold is unknown, but it is suspected that $n \geq 13$ suffices.

UNSOLVED PROBLEM 14   Flip Graph Diameter

Show that the diameter of the flip graph of a convex polygon with n vertices is $2n - 10$ for all n greater than 12.

UNSOLVED PROBLEM 15   Flip Graph Diameter Proof

Find a direct combinatorial proof of Theorem 3.42, one not using hyperbolic geometry.

Exercise 3.43. *Let P be a polygon with holes. Show that the flip graph of P is connected.*

Exercise 3.44. *Construct graphs with n nodes that have diameter 2. In general, can you construct graphs that have diameter k for an arbitrary value of k?*

Our adventure into the world of associahedra has thus far been combinatorial. We have considered these higher-dimensional polytopes as objects whose faces are labeled with diagonalizations of 2D polygons. In the early 1990s, a series of articles by Israel Gelfand, Mikhail Kapranov, and Andrei Zelevinsky placed the associahedron in the more general context of *secondary* polytopes. Although this development is beyond our scope here, we cannot resist sharing a glimpse of the field. The power of secondary polytopes derives from their providing associahedra a natural geometric world to inhabit. This takes us to a wonderful blend of combinatorics, geometry, and the convex hull.

Let P be a planar convex polygon with vertices $p_1, \ldots, p_n$. For a triangulation T of P, let

$$\phi_T(p_i) = \sum_{p_i \in \Delta \in T} \text{area}(\Delta)$$

be the sum of the areas of all triangles Δ incident to the vertex p_i. Figure 3.16 shows an octagon of total area 14, with three different triangulations. The numerical value within each triangle is the area of the triangle. The number $\phi_T(p_i)$ associated to each vertex p_i of the polygon is the sum of the areas of the triangles incident to p_i.

Let the *area vector* of a triangulation T be

$$\Phi(T) = (\phi_T(p_1), \ldots, \phi_T(p_n)).$$

So the area vector associates a value in $\mathbb{R}^n$ for each triangulation of the convex polygon. Of course, the number of potentially different area vectors associated to a convex polygon with n vertices is the Catalan number. The following result derives from the theory of secondary polytopes.

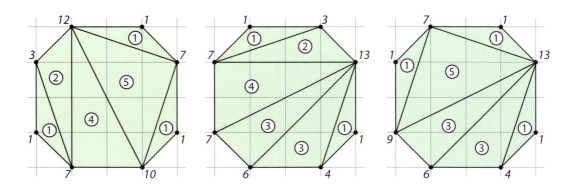

Figure 3.16. Area vectors may be read off counterclockwise around the boundary.

Theorem 3.45. *If P is a convex polygon with n vertices, the convex hull of the area vectors of all triangulations of P is combinatorially equivalent to the associahedron of dimension n − 3.*

Notice the difference in dimension. Let's consider a concrete example: a convex pentagon. Although the area vectors of the pentagon are sitting in the 5D space $\mathbb{R}^5$, the convex hull of these points results in the 2D associahedron. That is, the pentagon associahedron is lying in a 2D-plane that is sitting inside 5D space! By a similar argument, the area vectors of a convex hexagon are in $\mathbb{R}^6$, whose convex hull results in the 3D associahedron. Now that is elegant!

One final remark. The flip graph of a convex polygon does not depend on the shape of the polygon, but only on its number of vertices, because it records only the combinatorial structure of triangulations. Area vectors, however, do depend on the shape of the convex polygon, as well as on the combinatorial structure of diagonalizations. What is remarkable about Theorem 3.45 is that, although the convex hulls of the area vectors of different convex polygons are geometrically distinct, they have the same combinatorial structure for *any* convex polygon with *n* vertices.

Exercise 3.46. *Consider a regular pentagon having area 3. List the 5 area vectors for the 5 vertices of its flip graph (Figure 3.12). Argue that these points lie in a 2D subspace of $\mathbb{R}^5$, as per Theorem 3.45.*

3.4 DELAUNAY TRIANGULATIONS

We have so far studied relations among all the triangulations of a point set S as encapsulated in the flip graph: each node of the flip graph corresponds to a particular triangulation of S. There are many circumstances in which certain triangulations of S are valued more highly than others. An extremely important one is called the Delaunay triangulation, to which we now turn. This particular triangulation appears in numerous areas; it is most notably vital in the field of terrain reconstruction.

Most 3D maps of the Earth's surface are constructed starting from a finite sample S of points on the surface at which the surface height (altitude) has somehow been measured. From each of these points, a surface "terrain" is created that approximates the height of the nearby (unsampled) points. The method first considers the points S on the plane, constructs a triangulation of S on the plane, and then *lifts* each of the sample points to its correct height. This process lifts every triangle in the plane to a (generally tilted) triangle in 3D, providing a *piecewise-linear* terrain of the earth, to employ the technical term. Figure 3.17 provides an example of a terrain reconstruction.

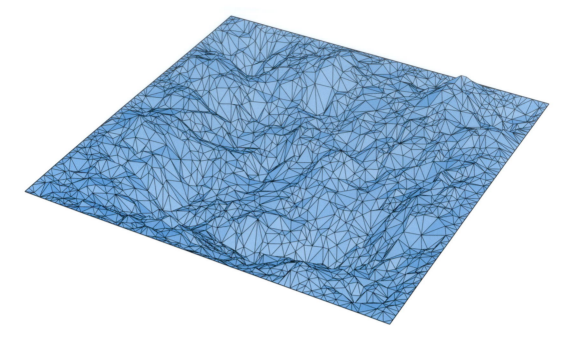

Figure 3.17. A piecewise-linear terrain reconstruction.

The natural question to ask is which of our triangulations is best for reconstructing a terrain from sampled heights. The true terrain of the earth is unknown except at the sample points. The choice of triangulation will have a major impact on the appearance of the terrain. In Figure 3.18, two identical point sets with heights are given, each with a different triangulation. Flipping the diagonal changes a deep valley to a steep mountain, making a big difference in the resulting terrain map.

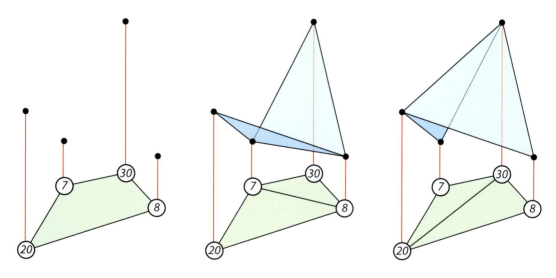

Figure 3.18. Mountains and valleys depend on the triangulation.

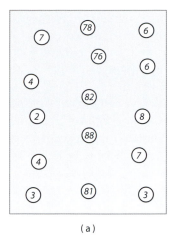 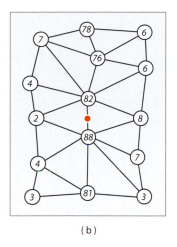 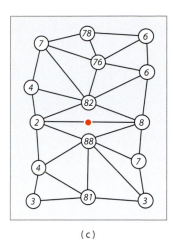

| (a) | (b) | (c) |

Figure 3.19. (a) Terrain point set, (b) one possible triangulation, and (c) another triangulation with two "skinny" triangles.

But how can we choose the appropriate triangulation if we have no further information? Our experience with natural terrains inculcates intuition that renders some terrains "more natural" than others to our eye. Let's rely on this intuition as we examine the sample terrain point set shown in Figure 3.19(a). From this information alone, the point set seems to have been taken from a long mountain ridge spanning north to south. Figure 3.19(b) shows a triangulation of this point set that captures this intuition of a mountain ridge. However, one flip (c) creates a deep valley that cuts the ridge in two. The marked red point in the center has height around 85 in part (b), but has a much lower height, about 5, in part (c).

What is it about the triangulation in Figure 3.19(b) that captures the terrain features so much more naturally than in (c)? The reason can be attributed to the fact the flipped edge in (c) produces *skinny triangles* compared to those in (b). In terrain reconstruction, the triangulations of choice are those which avoid skinny triangles by maximizing the smallest angle in any triangle over all triangulations. Indeed, the terrain in Figure 3.17 was constructed to maximize its angles in just this sense. We now show how to find this special triangulation.

Previously, we interpreted "general position" to mean that no three points are collinear or no two points share the same x-coordinate. In the following sections, the more relevant special (degenerate) circumstance is that when four points lie on the same circle. So *general position* will mean that no four points are cocircular.

Let T be a triangulation of our point set S, and suppose that T has n triangles. The *angle sequence*

$$(\alpha_1, \alpha_2, \ldots, \alpha_{3n})$$

of T is the list of all $3n$ angles of T sorted from smallest α_1 to largest α_{3n}. Using the angle sequence, we are now in a position to compare two different triangulations of S. Remember that by Theorem 3.13, the number of triangles of any triangulation of S is a constant; so all triangulations have angle sequences of the same length.

For two triangulations T_1 and T_2 of S, we say T_1 is *fatter than* T_2 (and write $T_1 > T_2$) if the angle sequence of T_1 is *lexicographically* greater than T_2. In other words, if $(\alpha_1, \ldots, \alpha_{3n})$ is the angle sequence for T_1 and $(\beta_1, \ldots, \beta_{3n})$ for T_2, then there is some $1 \leq k \leq 3n$ where $\alpha_i = \beta_i$ for all $i < k$ and $\alpha_k > \beta_k$. Thus

$$(20°, 30°, 45°, 65°, 70°, 130°)$$

is fatter than

$$(20°, 30°, 45°, 60°, 75°, 130°)$$

because $65° > 60°$ at the first position at which they differ, regardless of subsequent entries. The fattest triangulation is the one we seek, but how do we go about finding it? It turns out that there exists an elegant way to reach the fattest triangulation *via* edge flips. We begin with a definition.

Definition. Let e be an edge of a triangulation T_1, and let Q be the quadrilateral in T_1 formed by the two triangles having e as their common edge. If Q is convex, let T_2 be the triangulation after flipping edge e in T_1. We say e is a *legal edge* if $T_1 \geq T_2$ and e is an illegal edge if $T_1 < T_2$.

Notice that flipping one edge e alters six angles in the T_1 angle sequence, replacing them by six (in general) different angles in the T_2 sequence. So the effect of one flip is in general complex. But the definition just relies on the lexicographic ordering of the two triangulations, ignoring the details of how this ordering is achieved. It helps to complete this definition to declare that all the hull edges of a triangulation are legal. As our goal is to find the fattest triangulation, we are trying to avoid illegal edges in our triangulations.

Definition. For a point set S, a *Delaunay triangulation* of S, denoted $\mathrm{Del}(S)$, is a triangulation that only has legal edges.

This triangulation is named after Boris Delaunay, a Russian mathematician who lived from 1890 to 1980. It is not immediate that every point set has a Delaunay triangulation. Is it possible to remove all illegal edges in a triangulation without introducing new illegal edges during the removal process? The following algorithm answers YES and constructs the Delaunay triangulation in a remarkably simple manner.

> **EDGE FLIPPING**   Delaunay Triangulation Algorithm
>
> Let S be a point set in general position, with no four points cocircular. Start with any triangulation T. If T has an illegal edge, flip the edge and make it legal. Continue flipping illegal edges, moving through the flip graph of S in any order, until no more illegal edges remain.

Because illegal edges are being flipped, the angle sequence of the new triangulations strictly increases, and so the same triangulation can never reappear by this process. And since there are only a finite number of nodes in the flip graph, this algorithm must terminate.

By construction, the resulting Delaunay triangulation will be fatter than any of its neighbors in its flip graph. In other words, it will be a local fatness maximum. But this does not necessarily imply that it will be fattest over *all nodes* of the flip graph, the global maximum. We defer to the following chapter a proof that this is indeed the case: the Delaunay triangulation is the unique global maximum, the fattest among all triangulations of S.

Exercise 3.47. *Prove or disprove: under the algorithm above, it is possible for an edge to be legal and then later become illegal.*

★ **Exercise 3.48.** *Show that for every n, there is a triangulation of n points that requires $\Omega(n^2)$ flips to transform it into the Delaunay triangulation.*

In the Delaunay algorithm above, flipping one diagonal, as we mentioned before, involves twelve angles, all of which need to be compared in the angle sequences. We now show that there exists a far simpler test using circles based on an extension of a classical theorem[2] of planar geometry:

Theorem 3.49 (Thales). *For three points P, Q, and B on a circle, and A within and C outside the circle (see Figure 3.20), angle PAQ is greater than PBQ which is greater than PCQ.*

Sketch of Proof. Let O be the center of the circle. The three segments OP, OQ, and OB are radii and so equal in length. Thus triangles POB and QOB are both isosceles. From this it is not difficult to show that,

---

[2] Thales' theorem is often stated as the more specific claim that an angle inscribed in a semicircle is a right angle.

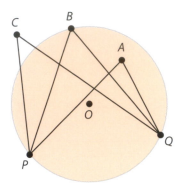

Figure 3.20. A diagram for Thales' theorem.

regardless of the position of B on the circle arc, angle POQ is twice angle PBQ. Indeed, this is a theorem of Euclid (*Elements*, Book III, Proposition 20): an angle inscribed in a circle is half the central angle subtending the same arc. This can be used to establish the claims that angle PAQ is larger, and PCQ smaller, than PBQ.   □

Exercise 3.50. *Provide a more formal proof of Thales' theorem.*

Consider the circumcircle of a triangle ABC with an additional point outside, on, or inside the circumcircle, respectively, as in Figure 3.21. The following proposition uses Thales' theorem to relate legal edges to circumcircles.

Proposition 3.51. *Let e be an edge of a triangulation, where $e = AC$ belongs to the two triangles ABC and ACD. Then e is a legal edge if D is outside the circumcircle of ABC and an illegal edge if D is inside the circumcircle.*

Recall that we have assumed that no four points are cocircular, so the case of D on the circumcircle does not arise.

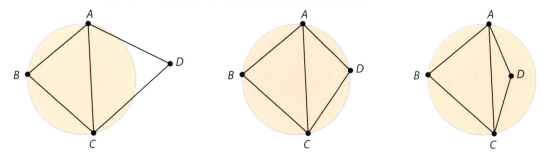

Figure 3.21. The circumcircle of a triangle ABC with an additional point outside, on, or inside the circumcircle, respectively.

Proof. Consider the left side of Figure 3.22, where D lies inside the circle defined by ABC. We show that AC is an illegal edge. Label the eight angles of the quadrilateral cut by both diagonals as on the right side of Figure 3.22. Because C is outside the circumcircle of ABD, by Thales' theorem, angle b_1 is larger than angle a_1. Similarly, because A is outside the circumcircle of BCD, by Thales again, angle b_2 is larger than angle a_2. Continuing in this manner, we see that $b_i > a_i$ for all i.

Moreover, since D is inside circle ABC, the angle sequence for edge AC must have $a_1, \ldots, a_4$ as its smallest angles. Thus for each of the six angles formed by edge BD, there exists a smaller angle formed by edge AC, proving that AC is illegal. A nearly identical proof works when D lies outside the circle. $\square$

Exercise 3.52. *Given the triangles defined in Proposition 3.51, show that D is outside the circumcircle of ABC if and only if B is outside the circumcircle of ACD. Prove this is true even if $ABCD$ does not form a convex quadrilateral.*

Based on Proposition 3.51, the following theorem classifies Delaunay triangulations in a novel and powerful manner:

Theorem 3.53 (Empty Circle Property). *Let S be a point set in general position, where no four points are cocircular. A triangulation T is a Delaunay triangulation if and only if no point from S is in the interior of any circumcircle of a triangle of T.*

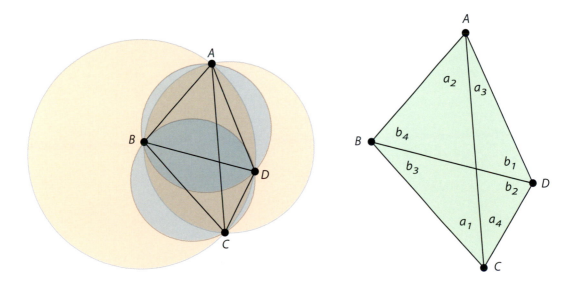

Figure 3.22. Legal edges and circumcircles.

Proof. If no point of S is interior to any of the circumcircles, then any flip will produce an illegal edge (by Thales' theorem). Thus all edges of the triangulation are legal.

We prove the converse of the statement using contradiction. Assume T is Delaunay and suppose there exist triangles whose circumcircles contain points in their interior. Such a situation would look like Figure 3.23(a) for a triangle ABC and point D within its circumcircle. Of all the triangles of T whose circumcircles contain points, choose the one whose point is closest to the edge of the triangle, that is, choose the one which minimizes the distance x given in part (b) of the figure.

Because T is Delaunay, all its edges are legal. So by Proposition 3.51, triangle BCD cannot exist in T. Let BCE be the triangle adjacent to ABC along edge BC. Again by the proposition, E must be outside the circumcircle of ABC, as in Figure 3.23(c). Notice that the circumcircle of BCE contains D, and D cannot be inside the triangle BCE. We have now reached a contradiction: D is a point inside the circumcircle of BCE, with the distance from D to EC less than the distance x.   □

Exercise 3.54. *Prove that the circumcircle of BCE contains D and that D cannot be inside the triangle BCE.*

Exercise 3.55. *Prove that the smallest angle of any triangulation of a convex polygon whose vertices lie on a circle is the same for each triangulation.*

Exercise 3.56. *For every n > 3, design a set of n points in the plane, no four cocircular, such that one vertex of the Delaunay triangulation has degree n − 1.*

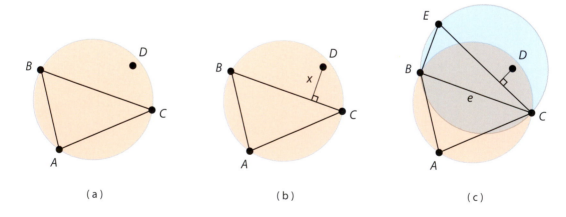

Figure 3.23. Proving the empty circle property.

3.5 SPECIAL TRIANGULATIONS

We close this chapter with a look at some special triangulations and their properties. So far we have considered one special node in our flip graph, the Delaunay triangulation. Another special node is the *minimum weight triangulation (MWT)*, defined as the one using the least amount of ink to draw it compared to all other triangulations. Each edge is viewed as having a *weight* equal to its Euclidean length. The minimum weight triangulation appears in numerous areas such as network topology, where it can be important to minimize the wiring cost of building a network infrastructure.

How can we go about searching for the minimum weight triangulation? Noting that skinny triangles usually have long edges and small angles, it might seem reasonable to guess that the Delaunay triangulation might indeed be the minimum weight triangulation. Unfortunately, we are not that lucky. Consider the point set with 33 points, where 32 of them are evenly spaced on a circle of radius 1, and the last point is at the center of the circle. Slightly perturb the points so that no four are cocircular.

For one triangulation, take the convex hull of the points and connect each of the hull points to the center point. Using the empty circle property (Theorem 3.53), it is not difficult to see that this is the Delaunay triangulation. A quarter of the circle illustrating this triangulation appears in Figure 3.24(a).

Exercise 3.57. *Prove that the triangulation in Figure 3.24(a) is Delaunay.*

For the second triangulation, again take the convex hull, connecting every adjacent point on the boundary. Now connect every second point on the boundary, making 16 new edges, and then add 8 new edges connecting every fourth point on the boundary. After connecting every eighth point with 4 new edges, we finish the triangulation by adding an

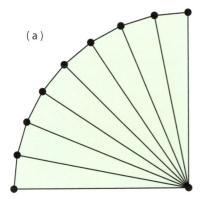

(a)

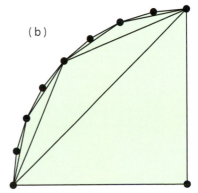

(b)

Figure 3.24. The minimum weight triangulation is not Delaunay.

edge from every eighth point to the central point. A quarter of the circle showing such a triangulation appears in Figure 3.24(b).

We now show that the second triangulation has a smaller total weight than the Delaunay triangulation. The total weight of the Delaunay triangulation is close to $2\pi + 32$, the circumference of the circle plus the 32 radii. The total weight for the second triangulation is far less than $8\pi + 4$, where 8π represents four layers around the circle and 4 is the length of the four edges to the center. Because $2\pi + 32 \approx 38 > 29 \approx 8\pi + 4$, we have an example where the Delaunay triangulation is not the minimum weight triangulation.

★ **Exercise 3.58.** *Is the second triangulation given in Figure 3.24(b) the minimum weight triangulation? If not, can you find another triangulation with smaller total weight?*

Now that we know the Delaunay triangulation is not always the minimum weight triangulation, the question arises: how can the minimum weight triangulation be found? One direct attempt is called the *greedy algorithm*: Given n points, consider all possible $\binom{n}{2}$ distances between them. Insert the edges one by one into the growing triangulation, choosing at each step the edge with smallest available length that does not cross a previously inserted edge. It was proved by Errol Lloyd that this *greedy triangulation* yields neither the Delaunay nor the minimum weight triangulation. In fact, it was a long-standing open problem to determine the computational complexity of finding the minimum weight triangulation. It was settled only in 2006 by Wolfgang Mulzer and Günter Rote, who showed that, alas, the problem is NP-hard.[3] Figure 3.25 shows a point set along with (a) the Delaunay triangulation, (b) the greedy triangulation, and (c) the minimum weight triangulation. The total weight of each triangulation is given, assuming the perimeter to be 100 units in length.

Instead of considering a complete triangulation of a point set, what if we were interested in just a *tree* that *spans* the point set? In other words, we want to draw edges using the least amount of ink such that all the points are connected to each other. Such an object is called the *minimum spanning tree (MST)* of a point set. The following remarkable theorem says that this tree is indeed a subset of the Delaunay triangulation, thus justifying the intuition that the Delaunay triangulation is weight-minimizing in some sense.

Theorem 3.59. *For a point set S, a minimum spanning tree of S is a subset of the Delaunay triangulation of S.*

---

[3] The *NP-hard* problems are at least as difficult as NP-complete problems, and perhaps worse. See the Appendix for further explanation.

279.6

274.9

274.1

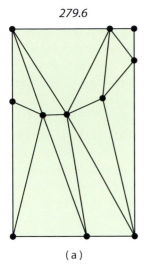 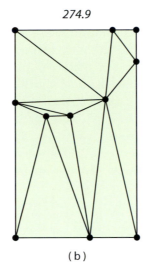 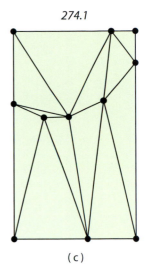

(a)   (b)   (c)

Figure 3.25. A point set along with its (a) Delaunay triangulation, (b) greedy algorithm triangulation, and (c) minimum weight triangulation. Figure courtesy of W. Mulzer and G. Rote.

Proof. Assume that edge AB is in the minimum spanning tree of S but not in the Delaunay triangulation. Consider the circle with diameter AB. Because AB is not a legal edge (by our definition of Delaunay triangulation), then by Proposition 3.51, there must be another point in this circle; call it C. Since AB is the diameter of the circle, $|AC| < |AB|$ and $|BC| < |AB|$.

Deleting AB from the minimum spanning tree will disconnect the tree into two trees, say T_A and T_B. The minimum spanning tree reaches (spans) all points of S, so C is in one of the two trees, say T_A. Now removing AB from the minimum spanning tree and adding BC will create a new spanning tree that is shorter in total length, yielding a contradiction.   □

UNSOLVED PROBLEM 16   **Minimum Weight**

Assume that the sum of the lengths of all the edges of the minimum weight triangulation is given. Find the minimum weight triangulation in polynomial time. (Even with the added information of the edge lengths, this may be difficult.)

So far in this chapter, we have concentrated on triangulations of one given point set. There are applications that naturally lead to comparisons of two triangulations of two different but related points sets X and Y, both with n points. For instance, motion-capture systems are used in filmmaking, recording the movements of actors dressed in special suits equipped with dozens of reflective markers. A computer then looks at the motion of all the markers and uses their movements to animate the character. (This method was heavily used to animate the Gollum character in Peter Jackson's *The Lord of the Rings* movie trilogy.) Here sets X and Y represent two time snapshots of the markers on the actor's suit with the triangulations providing interpolation guidance for the intermediate points. This is an instance where it would be useful to find a *compatible triangulation* of X and Y, first defined by Oswin Aichholzer, Franz Aurenhammer, and Hannes Krasser in 2001 in the following sense:

Definition. Given two planar point sets X and Y with n points each, along with triangulations T_X and T_Y, we say T_X and T_Y are *compatible* if there exists a bijection ϕ between the points of X and Y such that ABC is a triangle of T_X if and only if $\phi(A)\phi(B)\phi(C)$ is a triangle of T_Y.

Figure 3.26 gives an example of compatible triangulations. The two triangulations have the same *combinatorial structure* in terms of the gluing of their triangles, but they differ geometrically in angles and edge lengths.

Recall from Theorem 3.13 that the number of triangles in any triangulation with n points is $2n - 2 - h$, where h is the number of points on the hull. Therefore, it is a necessary condition for a compatible triangulation that the point sets have the same number of hull points. But is this condition sufficient? Astoundingly, this remains an open problem.

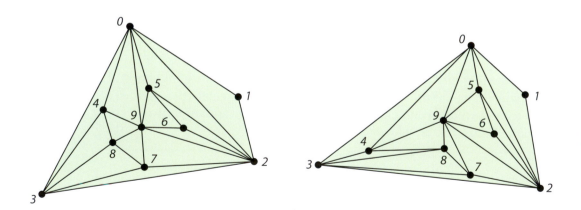

Figure 3.26. Two 10-point sets compatibly triangulated.

> **UNSOLVED PROBLEM 17**   **Compatible Triangulations**
>
> Does every pair of planar n-point sets with the same number of points h on the hull have a compatible triangulation? (This has been proven only for point sets with at most three interior points.)

Finding compatible triangulations involves finding a bijection between the points while simultaneously finding a triangulation. One cannot be done before the other. It was shown by Alan Saalfeld in 1987 that if the bijection between the points is fixed first, then compatible triangulations do not always exist.

Exercise 3.60. *Construct two n-point sets with the same number of points h on the hull. Provide a bijection between the point sets for which no compatible triangulations exist for this bijection.*

Exercise 3.61. *Given two polygons with n vertices, is it always possible to compatibly triangulate both polygons?*

The problem of finding compatible triangulations can be made easier if we are permitted to add extra points to the point sets, called *Steiner* points, named after the nineteenth-century Swiss mathematician Jakob Steiner. One might try methods of compatible triangulation using exterior Steiner points, that is, additional points added outside the convex hull of the original point set. However, this turns out to be trivial, as the following theorem shows:

Theorem 3.62. *Any two n-point sets S and T may be compatibly triangulated with the addition of two exterior Steiner points to each set.*

Proof. In the standard xy-coordinate system, order each set in terms of increasing y-coordinate, and when points have the same y-coordinate, order them in terms of decreasing x-coordinate. Explicitly,

$$S = \{p_1 = (x_1, y_1), \ldots, p_n = (x_n, y_n)\},$$

where if $i > j$, then either $y_i > y_j$ or $y_i = y_j$ and $x_i < x_j$. Similarly,

$$T = \{q_1 = (u_1, v_1), \ldots, q_n = (u_n, v_n)\},$$

where if $i > j$, then either $v_i > v_j$ or $v_i = v_j$ and $u_i < u_j$. Add a Steiner point p_L slightly below p_1 and far enough to the left of S so that edges $p_i p_{i+1}$ (between consecutive points of S) and $p_L p_i$ (between p_L and points of S) do not intersect pairwise. Similarly, add a Steiner point p_R slightly above p_n and to the right of S with the corresponding property.

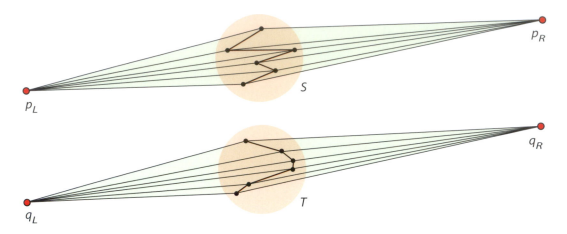

Figure 3.27. Example of sufficiency of two exterior Steiner points.

This ordering induces a vertical "zig-zag" as shown in Figure 3.27. Note that p_L and p_R exist because placing them arbitrarily far away yields edges arbitrarily close to horizontal. Add q_L and q_R to T in the same manner.

Let $S' = S \cup \{p_L, p_R\}$ and $T' = T \cup \{q_L, q_R\}$. By construction, the edges $p_L p_i$ and $p_i p_R$ (connecting each Steiner point to every point of the original set) together with the edges $p_i p_{i+1}$ yield a triangulation of S'; a similar construction gives a triangulation to T'. The bijection $f(p_*) = q_*$ shows these triangulations are compatible. □

The preceding theorem demonstrates the power of adding exterior Steiner points. But because the new Steiner points can be arbitrarily far away, the connection to the motivating applications is lost. A better solution might be to add Steiner points interior to the convex hull. This turns into a far more difficult problem, however. The best results so far produce compatible triangulations of two point sets after the introduction of $n - h - 3$ interior Steiner points, leaving much room for improvement.

We close this chapter on triangulations with objects that are not even triangulations! Nevertheless, they appear in numerous areas such as linkages and rigidity theory, which we discuss in Chapter 7. For example, they were the key to solving the art gallery problem with guards that cover $180°$ rather than $360°$.

Definition. A *pseudotriangle* is a polygon with exactly three convex vertices.

Figure 3.28 shows examples of several pseudotriangles. Instead of three straight sides connecting the three vertices, pseudotriangles have reflex

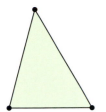 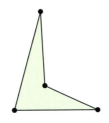 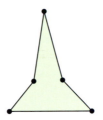 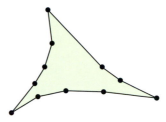

Figure 3.28. Examples of pseudotriangles.

chains (perhaps empty) connecting the three convex vertices. In particular, any triangle is a pseudotriangle.

Exercise 3.63. *Show that any polygon must have at least three convex vertices.*

Exercise 3.64. *Show that the convex hull of any pseudotriangle is a triangle.*

Just as a triangulation is a subdivision into triangles, a *pseudotriangulation* is a subdivision into pseudotriangles. Figure 3.29 shows several examples of pseudotriangulations of point sets.

Three more pseudotriangulations are given in Figure 3.30. There is a subtle difference between the first two and the last diagram. For the first two, notice that all the vertices have at least one surrounding angle greater than π. In the last diagram, all angles incident to the marked vertices are convex. This property is special enough to warrant a name.

Definition. A vertex is *pointed* if one of the angles it determines is reflex. A pseudotriangulation is pointed if all its vertices are pointed.

For any pseudotriangulation of a point set, all the hull vertices are clearly pointed because the exterior angle is reflex. This pointedness property can be noticed in Figure 3.29 as well. For the first two diagrams, the central vertex is incident only to convex vertices, whereas the other three pictures all provide pointed pseudotriangulations.

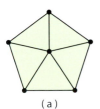 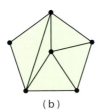 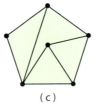 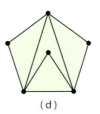 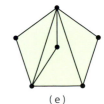

  (a)   (b)   (c)   (d)   (e)

Figure 3.29. Examples of pseudotriangulations, where (a)–(b) are also triangulations, with 5 triangles each, whereas (c)–(e) have 4 pseudotriangles each.

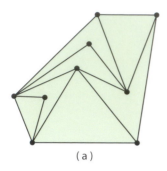
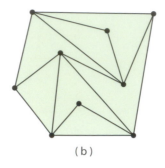
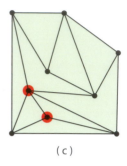

(a)   (b)   (c)

Figure 3.30. Examples of pseudotriangulations. All vertices are pointed except for the two marked ones in (c).

Theorem 3.13 provides a formula for the number of triangles in any triangulation of a point set S, dependent only on the number of points on the hull and in the interior of S. But this idea does not extend to pseudotriangulations, as shown by Figure 3.29: Here, although we are examining pseudotriangulations of the same point set, some have more pseudotriangles than others: two examples have 5 and three examples have 4. The key parameters revealing regularity are not hull versus interior points as for triangulations, but rather the number of pointed versus nonpointed vertices:

Theorem 3.65. *A pseudotriangulation of a point set S with p pointed vertices and q nonpointed vertices has $p + 2q - 2$ pseudotriangles and $2p + 3q - 3$ edges.*

Proof. Let t be the number of pseudotriangles and e the number of edges of the pseudotriangulation. By Euler's formula (Theorem 3.12), we obtain $(p + q) - e + (t + 1) = 2$, as there are t bounded faces and one unbounded one. Because each angle incident to a vertex is formed by two edges, the total number of angles equals $2e$. The number of reflex angles equals p, one at each pointed vertex, and the number of convex angles equals $3t$, one per pseudotriangle corner. Thus $2e = p + 3t$. This, along with the previous equation from Euler's formula, allows us to solve for e and t, establishing the claims.   □

The following is the analog of Theorem 3.13 for pointed pseudotriangulations.

Corollary 3.66. *A pointed pseudotriangulation of a point set S with n points has $n - 2$ pseudotriangles and $2n - 3$ edges.*

Exercise 3.67. *Find an algorithm which constructs a pointed pseudotriangulations for a given point set.*

Exercise 3.68. *For any point set, show that pointed pseudotriangulations have the least number of edges over all pseudotriangulations of the point set.*

The previous exercise is one reason pointed pseudotriangulations are sometimes called *minimal* pseudotriangulations. For generic pseudotriangulations, it is possible to remove some edges and still maintain a pseudotriangulation; see Figure 3.29(b) and Figure 3.30(c). However, pointed pseudotriangulations have the condition that removal of any edge loses the property of its being a pseudotriangulation.

The concept of a flip related two triangulations of a point set. Does this notion exist for pointed pseudotriangulations? The answer is YES due to the following result:

Theorem 3.69. *For a pointed pseudotriangulation T, there is exactly one flip for each interior edge e of T, that is, there exists a unique edge e′ such that removing e from T and replacing it with e′ is again a pointed pseudotriangulation.*

Exercise 3.70. *Show that deleting an edge shared by two pseudotriangles in a pointed pseudotriangulation produces a* pseudoquadrilateral, *that is, a (possibly degenerate) polygon with exactly four convex vertices.*

★ **Exercise 3.71.** *Prove Theorem 3.69 by showing that flipping an interior edge corresponds to the two ways of decomposing a pseudoquadrilateral into two pseudotriangles.*

In a triangulation, edges can only be flipped if their containing quadrilateral is convex. In contrast, *all* interior edges of a pointed pseudotriangulation can be flipped to form a different pointed pseudotriangulation on the same point set; see Figure 3.31 for some examples. Just as with

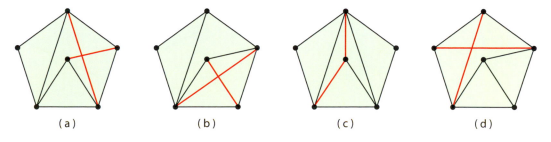

(a)   (b)   (c)   (d)

Figure 3.31. Each red edge of a pseudotriangulation can be flipped into the other.

triangulations, we can define a pseudotriangulations flip graph. It turns out that the flip graph of pseudotriangulations is also connected.

Exercise 3.72. *Prove that the flip graph of pseudotriangulations of a planar point set is connected.*

Recall from Section 3.3 that the associahedron is a polytope that captures the combinatorics of the flip graph of triangulations. In 2002, Günter Rote, Francisco Santos, and Ileana Streinu constructed a similar polytope that does the same for pointed pseudotriangulations. Figure 3.32(a) shows the flip graph of a point set using pointed pseudotriangulations and part (b) shows the three-dimensional polyhedron whose vertices and edges (its 1-skeleton) constitute the depicted flip graph. For points in convex position, note that pseudotriangulations coincide with triangulations. In this case, the pointed pseudotriangulation polytope constructed is exactly the associahedron.

> **UNSOLVED PROBLEM 18**   Pseudotriangulations
>
> For a planar point set S, is the number of pointed pseudotriangulations always at least the number of triangulations? This is conjectured to be true, with equality only when the points of S are in convex position. (This conjecture has been established for all points sets of 10 or fewer points.)

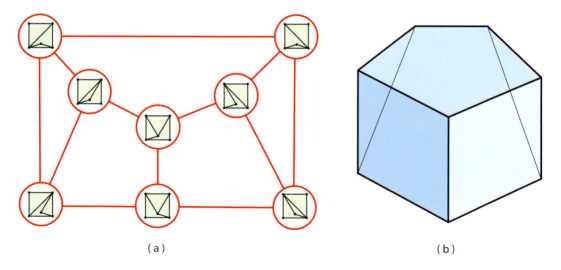

(a)   (b)

Figure 3.32. The (a) flip graph and (b) polyhedron of pointed pseudotriangulations of a point set.

SUGGESTED READINGS

Marshall Bern. Triangulations. In Jacob E. Goodman and Joseph O'Rourke, editors, *Handbook of Discrete and Computational Geometry*, chapter 25, pages 563–582. CRC Press LLC, 2nd edition, 2004.

A succinct and authoritative survey by one of the world's experts on triangulations, including what we call "tetrahedralization" of polyhedra, and partitioning into simplices in higher dimensions.

Herbert Edelsbrunner. *Geometry and Topology for Mesh Generation*. Cambridge University Press, 2001.

A gem of a book concentrating on 2D and 3D triangle and tetrahedral meshes, with Delaunay triangulations a central theme throughout the exposition. The proofs are concise, almost lapidary.

Charles Lawson. Transforming triangulations. *Discrete Mathematics*, Volume 3, pages 365–372, 1972.

A readable paper which first established the notion of a flip and proved that the flip graph of a planar point set is connected.

Daniel Sleator, Robert Tarjan, and William Thurston. Rotational distance, triangulations, and hyperbolic geometry. *Journal of the American Mathematical Society*, Volume 1, pages 647–681, 1988.

A remarkable and influential paper which looks at the diameter of the flip graph of a polygon (among other things) using combinatorics and the geometry of hyperbolic tetrahedra.

Jonathan Shewchuk. TRIANGLE: A Two-Dimensional Quality Mesh Generator and Delaunay Triangulator. `http://www.cs.cmu.edu/~quake/triangle.html`

Robust software to compute exact Delaunay triangulations and many variants.

Jon McCammond. Noncrossing partitions in surprising locations. *American Mathematical Monthly*, Volume 113, pages 598–610, 2006.

A geometric look at Catalan numbers, triangulations, and associahedra in several mathematics contexts. A nice survey of associahedra is given by Bill Casselman at `http://www.ams.org/featurecolumn/archive/associahedra.html`.

4 VORONOI DIAGRAMS

The convex hull captures the outer boundary of a point set. Triangulations partition the interior. In this chapter, the interest is in some sense on the points of the plane *not* in the given point set S. In particular, we study which point of S is closest to an arbitrary point not in S. This focus on "nearest neighbors" leads to the rich geometry of the Voronoi diagram (Section 4.1) and to the challenge of algorithmically constructing this diagram (Section 4.2). We will see there is an intimate connection via "duality" between Voronoi diagrams and the Delaunay triangulations of Chapter 3 (Section 4.3). And there is a beautiful and deep connection between both these structures and convex hulls in 3D (Section 4.4).

4.1 VORONOI GEOMETRY

In the context of Voronoi diagrams, the points of our given finite set S are often called *sites*. Imagine that the sites in S represent a local chain of post offices. If you lived somewhere in the plane, you would naturally want to go to the office closest to your home. So it is equally natural to associate with a post office p the region of points that are each closer to p than to any other site in S. The subdivision of the (infinite) plane into these regions is called the *Voronoi diagram* of the point set, with each region a *Voronoi region*. Figure 4.1 shows the Voronoi diagram for 10 post offices.

The manner in which Voronoi diagrams capture proximity makes them extremely useful in many practical applications, including pattern recognition, facility location, robot motion planning, cartography, and crystallography, just to name a few. They were first seriously studied by Georgy Voronoi in 1908, but have been rediscovered in many forms and so appear under various names, including Thiessen polygons and Dirichlet tessellations.

Notice that some regions in Figure 4.1 are bounded (e.g., the red region) and others are unbounded (e.g., the blue region). However, each region is convex in this example, regardless whether it is bounded or not. Will this always be true for any Voronoi region for any point set in the plane? What other properties do Voronoi diagrams possess? We now turn to answering such questions with a close consideration of their geometry. Let S be a collection of sites in the plane. The idea is to assign to each site

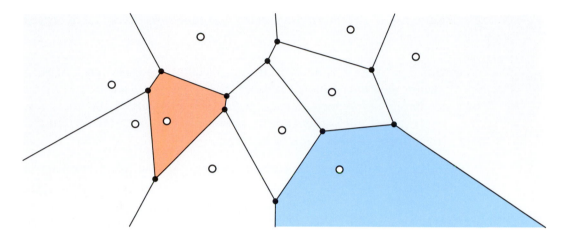

Figure 4.1. The Voronoi diagram for 10 sites.

the region that it influences.

Definition. The *Voronoi region* Vor(p) of a site p in S is

$$\text{Vor}(p) = \{x \in \mathbb{R}^2 \mid \|x - p\| \leq \|x - q\| \text{ for all sites } q \text{ in } S\},$$

where $\|p - q\|$ denotes the (Euclidean) distance between points p and q in the plane.

In words, Vor(p) is the set of all the points x that are at least as close to p as to any other site q in S. The points that lie on the boundary between regions do not have a unique nearest site. The *Voronoi diagram* Vor(S) is the collection of these boundaries: the set of all points in the plane that have more than one nearest neighbor. In Figure 4.1, the Voronoi diagram corresponds to the *Voronoi edges* and *Voronoi vertices* that partition the plane into the regions.

Let's look at the situation from the perspective of a single site p in S. If there is only one other site q in S, then the Voronoi diagram will simply be the perpendicular bisector of the segment pq. This bisector cuts the plane into two regions where the Voronoi region of p is the *halfplane* of points that contains p:

$$H(p, q) = \{x \in \mathbb{R}^2 \mid \|x - p\| \leq \|x - q\|\}.$$

If S has numerous sites, then we need to compare the distances of points between p and all other sites of S. This yields the following result:

Theorem 4.1. *The Voronoi region* Vor(p) *is the intersection of all halfplanes* $H(p, q)$, *where q is any other site in S.*

One of the fundamental results of discrete geometry is as follows:

Theorem 4.2. *The intersection of any (not necessarily finite) set of convex objects is convex.*

Proof. Let $\{X_i \mid i \in I\}$ be an arbitrary collection of convex sets, and let X denote their intersection. Consider arbitrary points p and q in X. By the definition of *intersection*, these points lie in every set X_i. Because every X_i is convex, the entire segment pq lies in every set X_i and therefore in their intersection X. Thus X is convex.   □

Since all halfplanes are convex regions, the following corollary is immediate:

Corollary 4.3. *All Voronoi regions are convex.*

Let's now turn our focus to the Voronoi vertices. When there are just three sites p, q, r in the point set S, the Voronoi diagram is formed by three perpendicular bisectors of the segments pq, pr, qr. Is it possible that these bisectors do not necessarily meet at a point, as in Figure 4.2(a)? The answer NO follows from a theorem of Euclid (*Elements*, Book IV, Proposition 5). Euclid proved that the perpendicular bisectors of the sides of a triangle—pqr in our case, shown in part (b)—must all pass through one point. In fact, this point, a Voronoi vertex, is the center of the (unique) circumcircle that passes through the triangle's vertices, as illustrated in (c).

Exercise 4.4. *Construct a point set with three sites whose Voronoi vertex is exterior to the triangle determined by the sites.*

Exercise 4.5. *Without invoking Euclid, provide a simple proof that Figure 4.2(a) is impossible.*

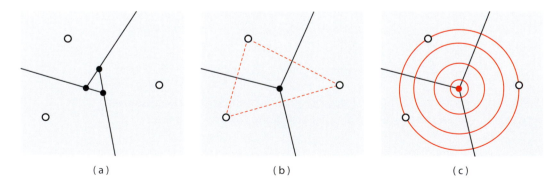

(a)   (b)   (c)

Figure 4.2. Bisectors and circumcircles with three sites.

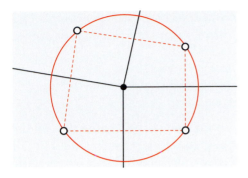 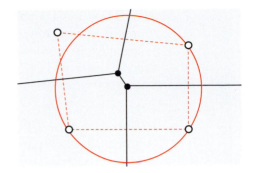

Figure 4.3. The Voronoi diagrams for four points.

What happens with four or more sites? Consider the two diagrams in Figure 4.3. The left one shows the case when four sites of the point set happen to lie on the same circumcircle, when the Voronoi vertex has degree 4. However, if one of the sites (such as the top left one) is moved slightly, as in the right figure, then the degree-4 Voronoi vertex splits into two degree-3 vertices. Thus the picture on the right is, in some sense, "generic" whereas the one on the left is degenerate. In this chapter, we consider a point set to be in *general position* if no four sites are cocircular, in which case all Voronoi vertices have degree 3.

As we have seen, intersections of bisectors create Voronoi vertices. However, certainly not *all* of these intersections become Voronoi vertices. Is there an easy way to find out which points on the plane will become Voronoi vertices? The following theorem answers this in the affirmative:

Theorem 4.6. *Let S be a point set with Voronoi diagram* Vor(S). *A point v is a Voronoi vertex of* Vor(S) *if and only if there exists a circle centered at v with three or more sites on its boundary and none in its interior.*

Proof. If v is a Voronoi vertex, then it must be incident to at least three Voronoi regions, say Vor(p), Vor(q), Vor(r). This implies that v must be equidistant from the three sites p, q, r, and hence there exists a circle centered at v with these sites on its boundary. If another site is inside this circle, then it would be closer to v, implying that the regions Vor(p), Vor(q), Vor(r) would not meet at v. This proves one direction of the claim.

Now assume such a circle centered at v exists, with at least three sites p, q, r on its boundary. Since the interior of the circle is empty, v must be on the boundary of each of the regions Vor(p), Vor(q), Vor(r). Hence, v is a Voronoi vertex. □

Exercise 4.7. *Let S contain the sites* $\{(1, 3), (1, 9), (1, 11), (3, 6), (4, 9), (6, 6)\}$. *Draw the Voronoi diagram of S.*

Exercise 4.8. *For each $n \geq 3$, is it possible to construct an example of a point set with n sites having no Voronoi vertices? How about having exactly one Voronoi vertex?*

Let's now turn to understanding Voronoi edges. We know that all the Voronoi edges are parts of perpendicular bisectors between sites, but not all these bisectors becomes Voronoi edges. We now present a geometric feature that characterizes the Voronoi edges, analogous to the theorem above:

Theorem 4.9. *Let S be a point set with Voronoi diagram* Vor(S), *and let e be a connected subset of the bisector between sites p and q of S. Then e is a Voronoi edge of* Vor(S) *if and only if for each point x in e, the circle centered at x through p and q contains no other sites of S in its interior or boundary.*

Proof. Suppose x is a point on the Voronoi edge between sites p and q. If the circle centered at x with p and q on its boundary contains another site r, then x would be in Vor(r) as well. Since this is a contradiction to being on a Voronoi edge, this circle is empty of other sites.

Now assume there exists an empty circle through p and q (and not through any other site) with x as its center. We then know that $\|x - p\| = \|x - q\|$ and $\|x - p\| \leq \|x - r\|$ for any other site r of S. Therefore, x must lie on the Voronoi diagram Vor(S) as either an edge or a vertex. By Theorem 4.6, x cannot be a Voronoi vertex, and thus x lies on a Voronoi edge. $\square$

We see in Figure 4.1 that two types of Voronoi edges are present: finite line segments (with two endpoints) and halflines or rays (with one endpoint). The only other type of edge is an infinite line (having no endpoints), as we saw when the point set S consisted of only two sites. The following theorem classifies the infinite-line edges.

Theorem 4.10. *A Voronoi diagram* Vor(S) *for a point set S has infinite-line edges if and only if all sites of S are collinear.*

Proof. If all sites of S are collinear, it easily follows that Vor(S) consists of parallel lines and no vertices. So we can assume that not all sites of S are collinear. Assume Vor(S) has an entire line called L, defined as the border of regions Vor(p) and Vor(q). Let r be a site in S not collinear with p and q, and (without loss of generality) assume that r is in $H(p, q)$. Then the perpendicular bisector of pr must intersect L. Moreover, the halfline $L \cap H(r, p)$ is not in Vor(S) because it is closer to r than to p. So the entire line L cannot be in Vor(S). $\square$

Corollary 4.11. *For a point set S, the Voronoi diagram* Vor(S) *is disconnected if and only if all sites of S are collinear.*

Proof. If Vor(S) is disconnected, then there must be a Voronoi region Vor(p) of S that separates the plane into two disjoint regions. Since Vor(p) must be a convex region by Corollary 4.3, the boundary of Vor(p) must be two parallel lines. By the theorem above, the points of S must be collinear.   □

We close this section with an enumeration result, bounding the total number of Voronoi vertices and edges, and showing the combinatorial complexity of the diagram to be $O(n)$.

Theorem 4.12. *Let S be a point set with $n \geq 3$ sites. Then* Vor(S) *has at most $2n - 5$ Voronoi vertices and $3n - 6$ Voronoi edges.*

Proof. Start by bending all the halflines of Vor(S) to meet an extra vertex located inside one of the unbounded regions, as illustrated in Figure 4.4. The result is a planar graph G, where each region of the graph is in one-to-one correspondence with a site of S; so the number of nodes of G is n, the number of sites in S. Also, Vor(S) and G have the same number of edges (call it e). And if Vor(S) has v vertices, G has $v + 1$ vertices.

Because G is planar, we may apply Euler's formula (Theorem 3.12), yielding the relationship $(v + 1) - e + n = 2$. Because summing the degrees of G counts each edge of G twice, and because each node of G has degree at least 3 (exactly 3 if S is in general position), we have

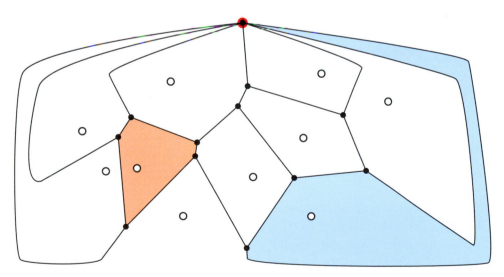

Figure 4.4. Adding an extra vertex and connecting to it the unbounded edges of Figure 4.1.

$3(v+1) \leq 2e$. Substituting $(v+1) = 2+e-n$ into this inequality yields the claimed bound on edges, and substituting $e = (v+1)+b-2$ yields the claimed bound on vertices.   □

This enumeration result implies that the number of vertices and edges is linear in the number of sites n. Naively, because there are $\binom{n}{2}$ site-site bisectors, the relationship could have been quadratic. The linear dependence on n means that many algorithms based on the Voronoi diagram can run quickly.

Exercise 4.13. *Describe the structure of the Voronoi diagram for the vertices of a regular polygon.*

Exercise 4.14. *For any point set S, prove that $\text{Vor}(p)$ is an unbounded region in the plane if and only if p is on the hull of S.*

Exercise 4.15. *For any point set S, prove that the average number of vertices of a Voronoi region of S is less than 6.*

4.2 ALGORITHMS TO CONSTRUCT THE DIAGRAM

Because of its numerous applications, considerable effort has been invested in designing algorithms to construct the Voronoi diagram. The most direct approach would be to find each Voronoi region separately as an intersection of $n-1$ halfplanes, using Theorem 4.1. Such a construction would yield a computational time of $O(n^2 \log n)$. Michael Shamos and Dan Hoey in 1975 provided a divide-and-conquer algorithm with time complexity $O(n \log n)$, which we will see (in Section 4.3) is optimal. However, this algorithm has proved difficult to implement, requiring careful attention to data structures.

In 1985, Steve Fortune discovered a clever algorithm with the same $O(n \log n)$ time complexity, but following a different paradigm known as *plane sweep*. The algorithm sweeps a vertical line left-to-right over the point set S, maintaining at all times the Voronoi diagram of the already swept points to the left. Although this is the general idea, the exact Voronoi diagram to the left depends on some sites to the right of the sweepline. So the algorithm is more subtle, maintaining a "parabolic front" that lags the sweepline slightly, with the property that the exact Voronoi diagram is constructed left of the parabolic front.

Of the many possible algorithms, we choose to explore a simple incremental algorithm for constructing the Voronoi diagram, described by Peter Green and Robin Sibson in 1977. It is a clean and elegant algorithm, and perhaps remains the most popular despite its $O(n^2)$ time complexity. The basic idea is similar to the incremental method used for convex hulls and triangulations. Assume that we have already built the

Voronoi diagram for k sites $\{p_1, p_2, \ldots, p_k\}$. We add a new site p to the plane, and so we need to convert the current Voronoi diagram to include the region Vor(p). Figure 4.5 shows this algorithm in action, starting with adding the extra site p into a previously constructed diagram and finishing with deleting the subdiagram inside Vor(p).

First we find the site, say p_1, whose Voronoi region Vor(p_1) contains p. (We relegate to an exercise the possibility that p lands on an edge or vertex of the diagram.) Such a site must exist because the Voronoi diagram partitions the plane. Now consider the perpendicular bisector of the segment $p_1 p$. Because Corollary 4.3 guarantees the regions to be convex, this bisector intersects the boundary of Vor(p_1) at exactly two points. Denote these points as x_1 and x_2 in such a way that the triangle $p x_1 x_2$ is counterclockwise. The line segment $x_1 x_2$ cuts the Voronoi region Vor(p_1) into two parts, one of which belongs to the region Vor(p) we are in the process of constructing; see the left side of Figure 4.6.

We now use the edge $x_1 x_2$ to obtain the remainder of the boundary of Vor(p). The point x_2 lies on the Voronoi edge between Vor(p_1) and its adjacent Voronoi region, say Vor(p_2). Now consider the perpendicular bisector of the segment $p_2 p$, which intersects the boundary of Vor(p_2) at two points, x_2 and another point, say x_3. The line segment $x_2 x_3$ cuts the Voronoi region Vor(p_2) into two parts, one of which belongs to the region Vor(p); see the right side of Figure 4.6. We repeat this procedure, finding a sequence of line segments enclosing p, until we return to x_1 again. These segments create a counterclockwise boundary of Vor(p).

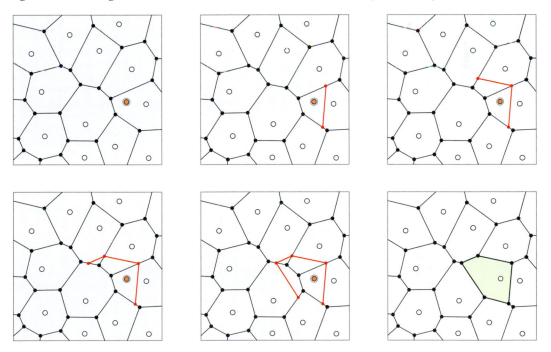

Figure 4.5. The incremental algorithm for creating Voronoi diagrams.

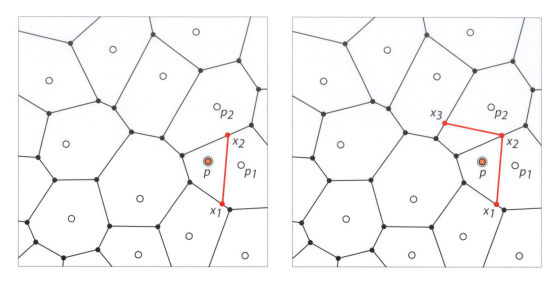

Figure 4.6. Labeled diagrams of the incremental algorithm in action.

Finally, we discard the old subdiagram inside Vor(p) to obtain the new Voronoi diagram. Notice that this procedure is entirely local, restricted to just one area of the Voronoi diagram. A summary of the procedure follows.

INCREMENTAL   Voronoi Diagram Algorithm $O(n^2)$

Given a constructed Voronoi diagram, find the region, say Vor(p_1), which contains the new site p. Draw the line segment x_1x_2 that is the perpendicular bisector to p_1p. Continuing from x_2, construct Vor(p) segment by segment until it closes up back at x_1. Remove the subdiagram inside this polygonal region to obtain the new Voronoi diagram, now containing p.

Exercise 4.16. *We claimed above that the changes to the Voronoi diagram are "local." Construct an example (for arbitrary n) in which every Voronoi region is altered by the addition of a new site, thus showing that the algorithm might need quadratic time in n.*

Exercise 4.17. *If the new added site p is outside the convex hull of $\{p_1, p_2, \ldots, p_k\}$, then Exercise 4.14 says that Vor(p) will be an unbounded region. Extend the algorithm to handle this situation.*

Exercise 4.18. *Extend the algorithm to handle the situation when the new added site falls not inside a Voronoi region but directly on a Voronoi edge or Voronoi vertex.*

Exercise 4.19. *Detail the geometric properties of a one-dimensional Voronoi diagram: n sites on a line. Design an algorithm to compute it and analyze its computational complexity.*

★ **Exercise 4.20.** *Given n − 1 points on a line, describe conditions on the points that determine whether they represent the one-dimensional Voronoi diagram of n sites on that line.*

★ **Exercise 4.21.** *Sketch a divide-and-conquer algorithm to construct Voronoi diagrams.*

Exercise 4.22. *The computation of the Voronoi diagram is so useful and widely needed that it is now included in standard computation libraries, such as* MATHEMATICA *or* MATLAB. *Learn how to compute the Voronoi diagram in one of these packages.*

4.3 DUALITY AND THE DELAUNAY TRIANGULATION

We mentioned that Voronoi diagrams encode proximity of points to sites. Site-to-site proximity may be captured by Voronoi region adjacency. A convenient representation of this adjacency relationship is the *dual graph* to the Voronoi diagram Vor(S). The nodes of the dual graph are the sites of S, and two sites are connected by an arc if they share a Voronoi edge. Figure 4.7 shows the graph dual to the Voronoi diagram of Figure 4.1 above.

In this figure, the arc between two sites is drawn to cross over its corresponding Voronoi edge. Because the Voronoi diagram is drawn on the plane, it follows that the dual graph is a planar graph, whose

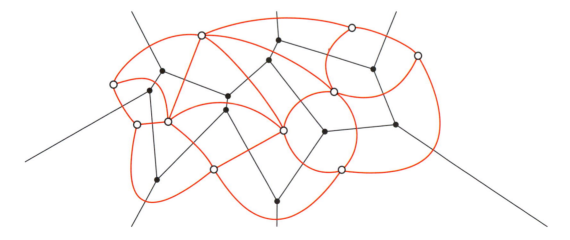

Figure 4.7. The dual graph of the Voronoi diagram.

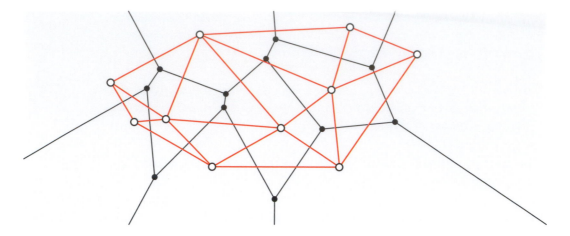

Figure 4.8. The straight-line dual graph of the Voronoi diagram.

arcs intersect only at the sites. The previous chapter focused on such connections between sites of S, resulting in triangulations. What would happen if we straightened the arcs of the dual graph, so the arc connecting sites p and q is replaced by the line segment pq? Figure 4.8 shows such a straightening for Figure 4.7.

In this figure, the result is a plane graph, where no two dual edges intersect. Would this be the case in general? In other words, is it always the case that the straightened arcs avoid crossing one another? The theorem below proved by Delaunay answers this in the positive. We first need the following technical lemma to prove the theorem.

Lemma 4.23. *Let A and B be two circles with chords that properly cross. Then at least one endpoint of one circle's chord is strictly inside the other circle.*

Proof. If one circle is contained within the other, the claim follows trivially. Otherwise, place the centers of A and B on a horizontal, with the leftmost point of A left of the leftmost point of B. Each circle naturally partitions the other into two parts: B cuts A into A_0 (the part outside B) and A_1 (the part inside B); similarly, A cuts B into B_0 (outside A) and B_1 (inside A).

Let $a_1 a_2$ be the chord of A with $a_1 \leq_x a_2$, and $b_1 b_2$ the chord of B with $b_1 \leq_x b_2$. Let L be the vertical segment determined by the two intersection points of $A \cap B$. If both a_1 and a_2 are in A_0, then $a_1, a_2 <_x L$, and if both b_1 and b_2 are in B_0, then $b_1, b_2 >_x L$. So if both these conditions hold, the chords cannot properly cross, as shown on the left of Figure 4.9. Therefore we must have either a_1 or a_2 in A_1, or we must have b_1 or b_2 in B_1. Any of the four possibilities results in some chord endpoint inside the other disk, as illustrated on

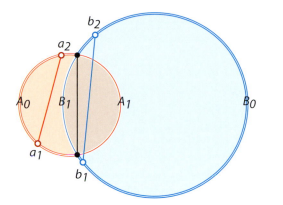 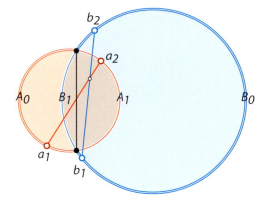

Figure 4.9. Crossing chords.

the right of Figure 4.9. Proper crossing forces some endpoint strictly inside, establishing the claim of the lemma.   □

Theorem 4.24. *The straight-line dual graph of* Vor(S) *is planar.*

Proof. We prove this by contradiction. Assume there are two edges $p_1 p_2$ and $p_3 p_4$ of the straight-line dual graph of Vor(S) that intersect. By Theorem 4.9, there exists a point x on the Voronoi edge between p_1 and p_2 such that the circle C_x centered at x and through p_1 and p_2 is empty. A similar empty circle C_y centered at y exists for p_3 and p_4. But Lemma 4.23 above showed that these conditions imply that either C_x or C_y is not empty, a contradiction.   □

Corollary 4.25. *If S is a point set in general position, with no four cocircular sites, the straight-line dual graph of* Vor(S) *is a triangulation of S.*

Proof. Because no four sites are cocircular, each Voronoi vertex has degree 3 and so corresponds to a triangle in the dual; see Figure 4.2(b) for a visual aid.   □

Following Corollary 4.25, the straight-line dual graph of Vor(S) is called the *dual triangulation* (or sometimes just the *dual*) of the Voronoi diagram. The duality associates each Voronoi region to a vertex (the region Vor(p) is associated to p) and each Voronoi vertex to a triangle.

———————————————

The previous chapter introduced the flip graph of a point set S, a graph encapsulating all triangulations of S. It is natural to wonder where the dual triangulation of Vor(S) fits into the flip graph. The following theorem, a foundational result in computational geometry, answers this question.

Theorem 4.26. *Let S be a point set in general position, with no four cocircular sites. The dual triangulation of* Vor(S) *is the Delaunay triangulation* Del(S) *of S.*

Proof. Theorem 4.6 shows that the closed circumcircle of each triangle in the dual of Vor(S) has no sites in its interior. Moreover, because S is in general position, the circumcircle of any triangle in the dual of Vor(S) contains only the vertices of that triangle. Theorem 3.53 shows that such a triangulation of S is the Delaunay triangulation. □

Although we have presented the Delaunay triangulation and the Voronoi diagram as almost accidentally related, the historical development followed a direct line from the Voronoi diagram to the Delaunay triangulation (Delaunay was a Ph.D. student of Voronoi at Kiev University) to the properties of the Delaunay triangulation (e.g., that it is the fattest triangulation). The uniqueness of the Voronoi diagram for a point set leads immediately to the same conclusion for the Delaunay triangulation:

Corollary 4.27. *Let S be a point set in general position, with no four cocircular sites. Then* Del(S) *is unique.*

Here the general position assumption is only needed to ensure that Del(S) is a triangulation. This theorem also provides a new proof of Theorem 3.22 from Chapter 3:

Corollary 4.28. *The flip graph of a planar point set is connected.*

Proof. By the Delaunay triangulation edge flipping algorithm, any triangulation can be converted to a triangulation with all legal edges. In other words, by the theorem above, any triangulation can be made into the (unique) Delaunay triangulation. This shows that there exists a path in the flip graph between any node and the node corresponding to the Delaunay triangulation. □

Exercise 4.29. *A Pitteway* triangulation of S is one for which every point in each triangle of the triangulation has one of its three vertices as the nearest neighbor among all sites of S. Show that not every Delaunay triangulation is a Pitteway triangulation.

We close this section by considering another algorithm for the Delaunay triangulation. Previously, we constructed Del(S) by flipping illegal

 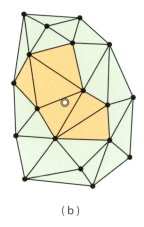 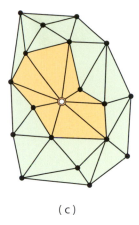

(a) (b) (c)

Figure 4.10. The dual graph to the Voronoi diagram.

edges of the triangulation. Now we use the duality between Voronoi and Delaunay to provide an incremental algorithm for building the Delaunay triangulation. Indeed, the one-to-one correspondence provided by duality implies that an algorithm for the Voronoi diagram must be an algorithm for the Delaunay triangulation "in disguise." As usual, we assume that no four sites of S are cocircular to avoid the distraction of degenerate cases.

Let S_k denote the first k sites of S and assume that we have already built the Delaunay triangulation $\text{Del}(S_k)$ of S_k. As a new site p is added to the plane, the current triangulation must be altered to include this site. We suppose the site p is inside the convex hull of the previous k sites, as shown in Figure 4.10(a). (We leave it as an exercise to consider the case when p is outside the convex hull.) By Theorem 3.53, a triangle t of $\text{Del}(S_k)$ will be affected if and only if the circumcircle of t contains p. We denote such triangles as *marked* triangles, shaded in Figure 4.10(b). Thus the changes made to $\text{Del}(S_k)$ are restricted to these marked triangles, each of whose circumcircles contains p.

The union of these marked triangles is a triangulation of a polygon inside the hull of the sites. Discard the diagonals of this polygon and add edges from p to each of the vertices of the polygon, as displayed in Figure 4.10(c). We claim that this is the Delaunay triangulation $\text{Del}(S_{k+1})$ of the sites, now with p included. It is clear by Theorem 3.53 that all the marked triangles must be changed. We thus need to show that all new triangles of the Delaunay triangulation must have p as a common vertex.

Suppose there is a new triangle t of $\text{Del}(S_{k+1})$ for which p is not one of its vertices. Then the circumcircle of t must be empty because it is a Delaunay triangle; in particular, p is not in the circumcircle. Thus

t cannot be a new triangle since it is a triangle of the original Delaunay triangulation $Del(S_k)$, leading to a contradiction.

INCREMENTAL   Delaunay Triangulation Algorithm $O(n^2)$

Given a constructed Delaunay triangulation $Del(S_k)$ with k sites, find the set of triangles of $Del(S_k)$ whose circumcircles contain the new site p. The union of these triangles is a triangulated polygon. Remove the diagonals of this polygon and add edges from p to each of the vertices of the polygon to obtain the new Delaunay triangulation, now containing p.

Notice that this algorithm runs in $O(n^2)$ time because it is just the dual of the Green and Sibson incremental Voronoi algorithm discussed earlier.

Exercise 4.30. *Extend the algorithm to handle the case when the additional site p is outside the convex hull of the previous k sites.*

Exercise 4.31. *Suppose we are given the Delaunay triangulation of a point set S with n points. Design an algorithm that constructs the Delaunay triangulation of the remaining $n - 1$ sites if a site from S is deleted. How does this algorithm change if the deleted site was on the hull of S or in the interior of $conv(S)$?*

Exercise 4.32. *What is the fewest number of triangles that will be altered by the addition of a site inside the hull? What is the maximum number of affected triangles?*

★ **Exercise 4.33.** *For a triangulation T of a point set S, let $d(T)$ be the diameter of largest circumcircle of any triangle in T. Prove that the Delaunay triangulation has the smallest value for $d(T)$ over all triangulations of S.*

UNSOLVED PROBLEM 19   Voronoi Diagram of Lines in 3D

What is the combinatorial complexity of the Voronoi diagram of a set of lines in 3D? Each Voronoi region includes the set of points closer to one of the given lines than to any other. There is a gap between a lower bound of $\Omega(n^2)$ and an upper bound that is essentially cubic.

4.4 CONVEX HULL REVISITED

We close this chapter with a remarkable connection between Delaunay triangulations and convex hulls in one higher dimension, discovered by Kevin Brown in 1979 and further developed by Herbert Edelsbrunner and Raimund Seidel in the early 1980s. The heart of this relationship involves the *paraboloid*

$$z = x^2 + y^2. \tag{4.1}$$

Let S be a point set in the xy-plane, with no four points cocircular. Associate to each site (x, y) a "terrain height" value of $x^2 + y^2$; this places the sites exactly onto the paraboloid. Now find the convex hull of this point set in $\mathbb{R}^3$. Discard the "top faces" of this hull, those faces which are visible looking straight down the z-axis from above. These faces are sometimes called the *upper convex hull*, whereas the remaining ones constitute the *lower convex hull*. The stunning result connecting convex hulls and Delaunay triangulations is given by the following theorem:

Theorem 4.34. *Given a point set S in the xy-plane, the Delaunay triangulation $\mathrm{Del}(S)$ is exactly the projection to the xy-plane of the lower convex hull of the points $(x, y, x^2 + y^2)$.*

Figure 4.11 shows 10 points on the plane whose triangulation derives from the projection of the lower convex hull of these points placed on

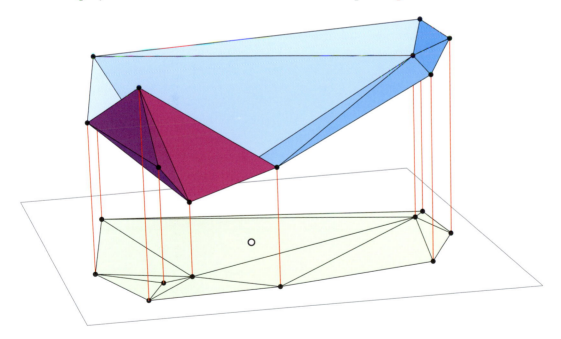

Figure 4.11. The Delaunay triangulation of a point set in the plane is obtained from the projection of the lower convex hull of these points on the paraboloid.

the paraboloid. The marked circle on the plane shows the origin in $\mathbb{R}^2$, the point at which the paraboloid touches the plane. By Theorem 4.34, the triangulation formed by this projection is the Delaunay triangulation of the point set.

Before proceeding to the proof, we pause to consider the geometry of the paraboloid. From multivariable calculus, the equation of the tangent plane at a point (a, b) is given by

$$z = 2ax + 2by - a^2 - b^2. \tag{4.2}$$

If this plane is shifted upward (in the z direction) by a distance of r^2, one obtains a new plane π given by

$$z = 2ax + 2by - a^2 - b^2 + r^2. \tag{4.3}$$

From conic geometry, this plane π intersects the paraboloid along an ellipse. The projection of this ellipse onto the xy-plane obtained by solving equations (4.1) and (4.3) yields

$$(x - a)^2 + (y - b)^2 = r^2 \, , \tag{4.4}$$

the equation of a circle. With this background, we begin the proof of the theorem.

Proof. Choose a face t of the lower convex hull, and let π be the plane defined by the three points of t on the paraboloid. We can shift this plane downwards (in the z direction) until it is tangent to the paraboloid. Let $(a, b, a^2 + b^2)$ be the point of tangency and let r^2 be the amount of downward shift on π. The equation of the plane π is then given by (4.3).

By the discussion above, the projection in the xy-plane of the three points defining t lies on a circle of radius r given by equation (4.4). Since t is a lower face of the convex hull, all other sites on the paraboloid lie above π, implying they project outside this circle of radius r. Therefore this circle determined by t is empty. By Theorem 3.53, the projection of t onto the xy-plane is a Delaunay triangle. Since this is true for all the lower convex hull faces, the full projection yields the Delaunay triangulation. □

A beautiful side observation comes from looking closer at Theorem 4.34. Note that this theorem holds independent of where the origin (and therefore the lowest point of the paraboloid) is with respect to the point set on the plane. Indeed, rotating and translating S will alter the lower convex hull but the projection of the triangulation remains the same! Figure 4.12 shows the same planar point set as Figure 4.11 but with a translation to the right. Again, the marked circle on the plane indicates

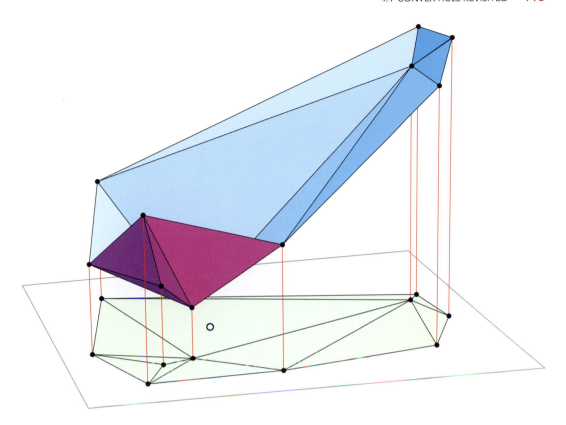

Figure 4.12. The point set in identical to that of Figure 4.11 but with a translation to the right. The lower hull yields a different shape but the planar triangulation is identical.

the origin in $\mathbb{R}^2$. Notice that the lower hull yields a different shape but the planar triangulation is still Delaunay.

We know from Chapter 2 that the convex hull of points in 3D can be obtained in $O(n \log n)$ time complexity. This implies that the Delaunay triangulation (as well as the Voronoi diagram) of points in $\mathbb{R}^2$ can be computed in the same time. One can show that this relationship between Delaunay and convex hulls holds in higher dimensions as well. Thus Delaunay tetrahedralizations, which are essential for solid meshing of 3D objects, can be obtained by constructing the 4D hull of the 3D points projected up to a paraboloid in $\mathbb{R}^4$.

We have so far emphasized how the lower hull of the paraboloid points project to the Delaunay triangulation. A dual view, which we only sketch, completes the picture. Imagine constructing all the planes tangent to the paraboloid over each site (a, b), following equation (4.2), and viewing these intersecting planes from $z = +\infty$. One sees the *upper envelope* of the planes — the pieces not obscured by any other plane. This upper envelope projects precisely to the Voronoi diagram!

The correspondence among these four structures — the Voronoi diagram, the Delaunay triangulation, the convex hull in 3D, and the envelope

of tangent planes — can be summarized briefly by the table below
The combinatorial duality between Voronoi diagrams and Delaunay
triangulations is a shadow (literally) of the projective duality between
lower hulls of points and upper envelopes of planes.

Voronoi	Delaunay	Lower Hull	Upper Envelope
site	site	lifted site	tangent plane to lifted site
Voronoi region	site	hull vertex	envelope face
Voronoi edge	Delaunay edge	hull edge	envelope edge
Voronoi vertex	Delaunay triangle	hull face	envelope vertex
3 coincident bisectors	empty circumcircle	supporting plane	3 coincident planes

This correspondence implies that an algorithm for one can be converted into an algorithm for the other. Thus the three incremental
algorithms we described for the 3D hull (Section 2.7), for the Delaunay
triangulation (Section 3.4), and for the Voronoi diagram (Section 4.2)
are all in some sense the same. Similarly, the Preparata-Hong divide-
and-conquer algorithm for the 3D hull "projects" to the Shamos-Hoey
divide-and-conquer algorithm for the Voronoi diagram, with the delicacy
of the former explaining the difficulty of the latter. The somewhat
mysterious parabolic fronts in Fortune's sweep algorithm are no longer
mysterious from the 3D hull point of view, and an implementation
of the algorithm in terms of Delaunay triangulations is conceptually
cleaner.

Exercise 4.35. *Extend the proof of Theorem 4.34 to relate the 3D
Delaunay tetrahedralizations and 4D convex hulls.*

★ **Exercise 4.36.** *Consider the upper convex hull, the set of faces we discarded. If the lower convex hull projects to form Del(S), to what object
does the upper convex hull project for S?*

SUGGESTED READINGS

Atsuyuki Okabe, Barry Boots, and Kokichi Sugihara. *Spatial Tessellations: Concepts and Applications of Voronoi Diagrams*. John Wiley & Sons, 2nd edition, 2000.
A complete and comprehensive reference on just about every nuance of Voronoi diagrams in 2D. It is especially valuable for connections to Geographic Information Systems (GIS).

Steve Fortune. Voronoi diagrams and Delaunay triangulations. In Jacob E. Goodman and Joseph O'Rourke, editors, *Handbook of Discrete and Computational Geometry*, chapter 22, pages 513–528. CRC Press LLC, 2nd edition, 2004.
A masterfully succinct survey by the inventor of Fortune's sweepline algorithm, originally described in "A sweepline algorithm for Voronoi diagrams" (*Algorithmica*, Volume 2, pages 153–174, 1987).

Franz Aurenhammer and Rolf Klein. Voronoi diagrams. In Jörg-Rüdiger Sack and Jorge Urrutia, editors, *Handbook of Computational Geometry*, chapter 5, pages 201–290, Elsevier, 2000.
Another masterful survey by leading researchers, including many applications, and 90 bibliographic references.

Mark de Berg, Marc van Kreveld, Mark Overmars, and Otfried Schwarzkopf. *Computational Geometry: Algorithms and Applications*. Springer-Verlag, 3rd edition, 2008.
A well-written textbook for advanced undergraduates and beginning graduates with a strong focus on algorithms in computer science. In particular, Chapter 7 contains a clear and detailed presentation of Voronoi diagrams and their generalizations.

Herbert Edelsbrunner and Raimund Seidel. Voronoi diagrams and arrangements. *Discrete and Computational Geometry*, Volume 1, pages 25–44, 1986.
The seminal paper that clarified the relationships between Voronoi diagrams, paraboloids, and arrangements. Builds on the connection between Voronoi diagrams and convex hulls first elucidated by Kevin Brown in his 1979 paper "Voronoi diagrams from convex hulls" (*Information Processing Letters*, Volume 9, pages 233–228).

5 CURVES

In this chapter, we extend the Voronoi diagram to apply to curves rather than to just sites, leading to two generalizations: the medial axis (Section 5.1) and the straight skeleton (Section 5.2). Both can be viewed as created by "offsetting" the polygon boundary, which leads to the Minkowski sum (Section 5.3) and convolution (Section 5.4). This in turn brings us naturally to curve shortening (Section 5.5), which connects to several deep theorems of mathematics, most notably the Poincaré conjecture using the heat equation (Section 5.6). Finally, we look at curve reconstruction (Section 5.7), an important practical task whose algorithms employ Voronoi diagrams, Delaunay triangulations, and the medial axis.

5.1 MEDIAL AXIS

In preparation for generalizing, let's review several equivalent definitions of the set of points that constitute the Voronoi diagram $Vor(S)$ of n sites S in the plane:

1. $Vor(S)$ is the locus[1] of centers of maximal empty disks: disks whose interior contain no points of S.
2. $Vor(S)$ is the locus of points to which there are two or more nearest sites.
3. $Vor(S)$ is the set of "quench points" if the plane is burned uniformly and simultaneously from every site in S.

Now, rather than a discrete set of sites S as the source of the Voronoi diagram, consider the boundary ∂P of a convex polygon P as the source. Then we seek an object that has the same three properties as $Vor(S)$. That object is known as the *medial axis* of P in the computer science literature, and the *cut locus* of ∂P in the mathematics literature; we will employ both names. Examples are shown in Figure 5.1.

Definition. The *medial axis* $M(P)$ of a polygon P (also known as the *cut locus* of ∂P) is the closure of the set of points in P that have two or more closest points among the points of ∂P.

---

[1] In geometry, a *locus* is a set of points satisfying some constraint or sharing some property.

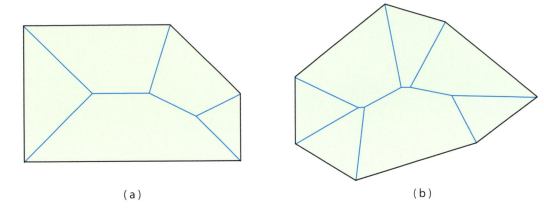

Figure 5.1. Convex polygons and their medial axes marked in blue.

Although this definition can be generalized considerably beyond polygons, for now we focus our attention on *convex* polygons. As might be guessed from Figure 5.1, the medial axis of a convex polygon is a *geometric tree* of straight segments whose leaves are the vertices of P. The reason for the definition stipulating "the closure" of the set of points is that, without that qualification, the vertices themselves would not be part of the medial axis, as each is its own unique closest point. Thus it is convenient to close the otherwise open set to incorporate the vertices.

The reason for the name "medial axis" is that it runs in some sense down the middle of the shape. The medial axis was introduced by Harry Blum in 1967 for studying biological shape and since then has found a wide variety of applications, several of which we will touch upon. The reason for the name "cut locus" is that straight paths from ∂P cease being shortest paths at the cut locus — they are "cut" there. The notion of cut locus was introduced in 1905 by Poincaré (whom we will encounter again in Section 5.5) and has become a standard object of study in Riemannian geometry.

Let's return to the three properties of Voronoi diagrams that we claimed the medial axis shares. Notice that points on the medial axis $M(P)$ of a polygon P are centers of *maximal disks* that touch ∂P in two or more distinct points. This is depicted in Figure 5.2, which reexamines the polygon of Figure 5.1(b) in detail. The maximal disks are shaded, having two or more distinct points of tangency with ∂P. For points on the interior of a segment of $M(P)$, as in Figure 5.2(a), there are two points that touch ∂P. And there are k points touching ∂P at a degree-k vertex of $M(P)$, as in Figure 5.2(b). For this reason, the radii from the center m of the disks to the touching points represent the two or more distinct shortest paths to ∂P.

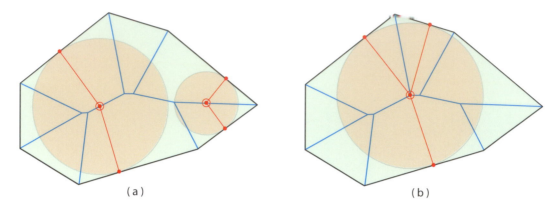

(a)   (b)

Figure 5.2. The maximal disks associated to (a) interior segments of $M(P)$ and (b) a degree-3 vertex of $M(P)$.

If the polygon encloses dry grass, starting a grassfire along the boundary marches in parallel from the boundary, with the "quench points" (where fire meets fire) along angle bisectors. Each quench point m in $M(P)$ may be associated with the time at which the fire reaches m, which is just the radius of the maximal disk centered there. Indeed, the map from P to $M(P)$ is sometimes known as the *grassfire transformation*.

Exercise 5.1. *Show that the medial axis of a convex polygon with n vertices could have a vertex of degree n.*

Exercise 5.2. *Compute and describe the medial axis for a rectangle.*

Exercise 5.3. *What is the maximum and minimum number of edges the medial axis tree M(P) can have for a convex polygon with n vertices?*

There is a natural polyhedron, the *medial axis polyhedron*, sometimes associated with P. Over each point m in $M(P)$, erect a vertical segment with height equal to the radius of the maximal disk centered on m. Now take the convex hull of P and all these vertical heights. The result is illustrated in Figure 5.3, which shows the medial axis polyhedron for the polygon given in Figure 5.1(b). Each face slants at $45°$ with respect to P because it represents the inward parallel uniform-rate march of the corresponding edge.

There is an attractive *physical* model of the medial axis. Imagine cutting out a polygon P from a thick piece of wood. Now pour dry sand on top of it, letting excess sand fall off the edges. The sand slopes in "facets," with all facets tilted at the same angle with respect to the polygon. Although this angle depends on the quality of the sand, the humidity, and other factors, it is not difficult to show that the resulting shape is that of the medial axis polyhedron. The "ridges" of the sand pile, viewed from above, form (an approximation to) the medial axis of P.

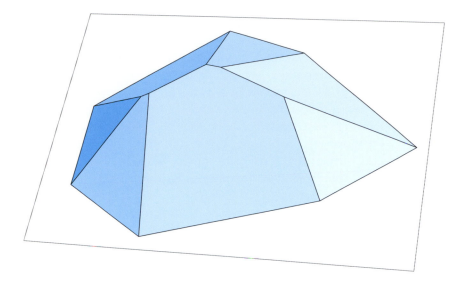

Figure 5.3. The medial axis polyhedron corresponding to Figure 5.1(b).

Figure 5.4 shows photographs of two examples of this construction. This model even works for nonconvex polygons, which we examine later.

Exercise 5.4. *Suppose the roof of an* L × W *rectangular building (assume* L ≥ W) *is a medial axis polyhedron. When rain falls on the roof, what percentage falls off of each edge as a function of L and W? Moreover, if friction causes raindrops to roll down the sloped roof at constant velocity v, what is the longest time a drop is on the roof before falling off, as a function of v, L, and W?*

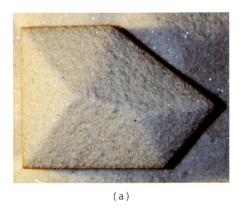

(a)

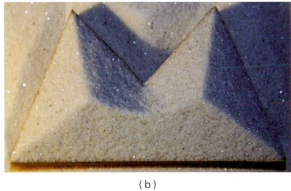

(b)

Figure 5.4. Physical models of the medial axis constructed using sand for (a) a convex irregular pentagon and (b) a nonconvex pentagon.

Let's look at the medial axis construction for convex polygons. Let $v_1, \ldots, v_n$ be the vertices and $e_i = v_i v_{i+1}$ the edges of P. It should be clear that the segment of $M(P)$ incident to v_i is the angle bisector at v_i, whose points are centers of disks touching e_{i-1} and e_i. What is perhaps less clear is what happens as these bisectors meet, that is, how the interior of the tree $M(P)$ is determined. There is an elegant recursive/inductive algorithm for constructing the medial axis of a convex polygon, which we now explain. We will use the convex heptagon (call it P_7) from Figure 5.1(b) to illustrate this construction.

Imagine the bisectors of each vertex angle growing inward during the grassfire transformation. At some time t, the first pair intersect. Perhaps not surprisingly, the first pair to intersect always constitute adjacent vertices (Exercise 5.5). In Figure 5.5(a), the bisectors from v_2 and v_3 of polygon P_7 are the first to meet, say at point x. This meeting constitutes (in general) a degree-3 node of $M(P)$, where the maximal disk touches three edges incident to the two vertices, in this case e_1, e_2, e_3.

As the fire continues to burn, it is the bisector of e_1 and e_3 that emerges from x — edge e_2 will no longer contribute to the construction of the diagram, its role having been exhausted. Thus we can extend e_1 and e_3 to meet at a new vertex $v_{1,3}$ of a polygon of one fewer vertex, P_6, as shown in Figure 5.5(b). The medial axis we seek, $M(P_7)$, is a slightly altered version of $M(P_6)$, which we do not yet know. The medial axis $M(P_6)$ can be derived by identifying v_6 and v_7 as the next bisectors to meet, and extending the two edges e_5 and e_7 to "engulf" e_6, as in Figure 5.5(c).

Continuing in this manner, we ultimately arrive at P_3, a triangle, whose medial axis is simply the three angle bisectors, which meet at the center of the inscribed circle. This is exactly what we saw in Figure 4.2, and we see it also in Figure 5.5(e). One can view this algorithm as constructing $M(P)$ in pieces top-down from $P = P_7$ as illustrated in Figure 5.5, or as a recursive algorithm that builds $M(P)$ bottom-up from P_3, as described below.

MEDIAL AXIS $O(n^2)$

Let P_n be an n-vertex convex polygon. Identify the two adjacent vertices v_i and v_{i+1} whose bisectors meet first, at point x. Extend edges e_{i-1} and e_{i+1} over e_i to meet at a new vertex v and call the resulting polygon P_{n-1}. Compute $M(P_{n-1})$ recursively and delete xv from $M(P_{n-1})$ and add xv_i and xv_{i+1} to form $M(P_n)$.

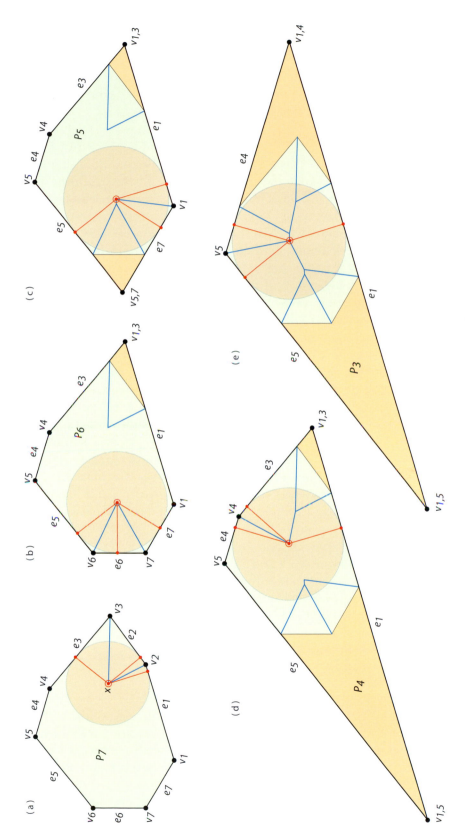

Figure 5.5. Construction sequence for the medial axis in Figure 5.1(b). At each step, two bisectors meet at the center of a maximal disk touching three consecutive edges, and then the middle edge is discarded to create a polygon with one less vertex.

Exercise 5.5. *Prove that the first two bisectors of a convex polygon P to meet are adjacent.*

Exercise 5.6. *Use this algorithm for the polygon in Figure 5.1(a) and draw the different stages of the construction.*

What can we say about the time complexity of this algorithm? As mentioned, the first two bisectors of P to meet must emanate from adjacent vertices. So we have $O(n)$ pairs of adjacent bisectors to compare. How can we determine which is the first pair to meet? Remember the time associated with a point m in $M(P)$ is the radius of the maximal disk centered at m. And notice that we know which edges are closest to m: For bisectors from v_i and v_{i+1}, the three edges tangent to the maximal disk centered at their intersection m are the three consecutive edges e_{i-1}, e_i, e_{i+1}. So with each pair of adjacent bisectors, we may compute the time-to-meet in constant time — as the radius of the circle centered at m that is tangent to those three edges. This permits us to compute a new polygon P_{n-1} by intersecting the lines containing e_{i-1} and e_{i+1}. So with $O(n)$ time of total work, we can recursively construct $M(P_{n-1})$ and from that derive $M(P_n)$. This leads to an $O(n^2)$ time algorithm. With a bit more attention to efficiency, the algorithm can be implemented to run in $O(n \log n)$ time.

Exercise 5.7. *Show how to implement the algorithm described for computing the medial axis of a convex polygon to run in $O(n \log n)$ time. (This exercise requires familiarity with "priority queue" data structures.)*

With considerably more work and cleverness, it is possible to compute the medial axis in $O(n)$ time, not only for convex polygons but for arbitrary polygons. Although we will not pursue the algorithmic aspects of this problem further, we next examine the geometric properties of the medial axis for nonconvex polygons.

Nonconvexities in the polygon P introduce a new element to the medial axis: Although $M(P)$ remains a tree whose leaves are the vertices, the edges of the tree associated with reflex vertices are parabolic arcs. Recall that a parabola is the locus of points equidistant from a point (the *focus*) and a line (the *directrix*). Consider the pentagon P on the left of Figure 5.6. Bisectors emerge from the four convex vertices, meet at a degree-3 node of $M(P)$ and continue as before, but those bisectors merge smoothly to an arc of a parabola with focus v and directrix the

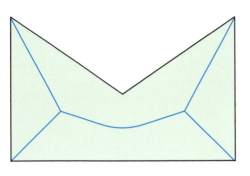 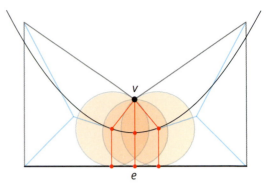

Figure 5.6. The central arc lies on the parabola determined by the vertex v and the edge e, where the maximal disks centered on that arc touch e and v.

line containing e, drawn on the right side of Figure 5.6. That arc contains points equidistant from v and e. The sand model previously shown in Figure 5.4(b) reproduces (approximations of) the parabolic arcs. The presence of parabolic arcs in the medial axis certainly complicates the computation for arbitrary polygons, but ultimately the same linear-time complexity can be achieved.

Exercise 5.8. *Is there a nonconvex polygon whose medial axis does* not *contain any parabolic arcs, but instead is composed entirely of straight segments?*

Exercise 5.9. *What is the minimum number of edges the medial axis tree $M(P)$ can have for an arbitrary polygon of n vertices?*

The medial axis has many applications, including shape recognition (especially biological shape), character recognition and font design, and GIS (Geographic Information Systems). The medial axis may be defined for 3D shapes as well (e.g., as the locus of the centers of maximal inscribed balls), and has been used for guiding NC (Numerically Controlled) machining, for partitioning 3D objects into pieces, and for 3D symmetry detection. The medial axis has also become an important theoretical tool in surface sampling theory and in the development of meshing software, a topic we cover in Section 5.7.

5.2 STRAIGHT SKELETON

Although the medial axis is both mathematically elegant and practically significant, the presence of parabolic arcs makes it problematical for some applications. A related alternative structure known as the *straight skeleton $S(P)$* for a polygon P was proposed in the mid-1990s and has since found many applications. For a convex polygon, the straight skeleton is identical to the medial axis, but for nonconvex polygons, the straight skeleton has no parabolic arcs. Compare the straight skeleton of Figure 5.7(a) with its medial axis in Figure 5.6. It is perhaps best explained via a variant of the grassfire transformation.

Imagine shrinking ∂P via a parallel translation of all edges at the same speed inward, with each vertex following the angle bisector. Reflex vertices also travel on angle bisectors, which implies that the incident edge grows in length at that endpoint. The shrinking continues until one of two events occurs, illustrated in Figure 5.7(b).

1. An edge shrinks to zero length. This is exactly the event we saw with the medial axis of convex polygons and, just as in that circumstance, the process continues with the new vertex tracking the bisector of the neighboring edges.
2. A reflex vertex collides with an edge. At this point, the original polygon is "pinched off," creating two new polygons (triangles in the figure). The shrinking process then continues on the two polygons independently.

Shrinking stops on a subpolygon when its area reduces to zero.

Like the medial axis, $S(P)$ is a tree, this time of straight segments whose leaves are the vertices of P. The segments are straight because they are always pieces of angle bisectors. Unlike the medial axis, it partitions the interior of the polygon into n regions, precisely one per edge, a useful partition property that has been applied to the mathematics of origami. Figure 5.7 shows $S(P)$ partitioning the pentagon P into five regions, whereas Figure 5.6 shows that $M(P)$ delimits four regions.

Exercise 5.10. *Construct the medial axis and the straight skeleton of Figure 1.8(a).*

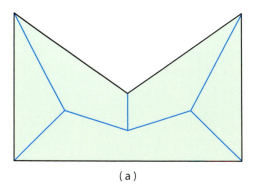
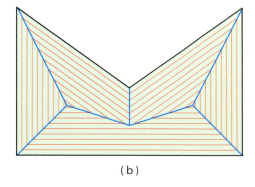

(a)   (b)

Figure 5.7. (a) The straight skeleton of the nonconvex polygon in Figure 5.6, along with (b) the shrinking process used for its construction.

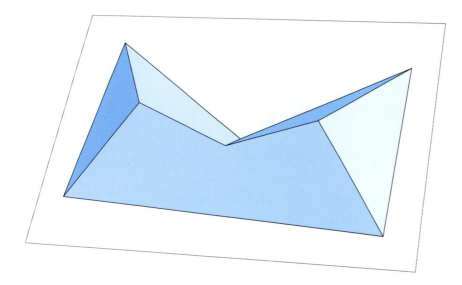

Figure 5.8. The straight skeleton polyhedron of Figure 5.7.

One of the most interesting applications of the straight skeleton derives from the *straight skeleton polyhedron*. Like its medial axis polyhedron counterpart, this polyhedron has constant-slope facets over each face of the partition of P induced by $S(P)$. Unlike the curved medial axis surface obtained for nonconvex P, the straight skeleton polyhedron facets are all flat. Figure 5.8 depicts the straight skeleton polyhedron corresponding to Figure 5.7. One can view this polyhedron as the solution of how to construct a piecewise flat roof over the walls of ∂P with all sections sloped at the same angle. This polyhedron has the attractive property that rainwater runs off each roof facet to the edge of ∂P that "generates" the facet. Thus one could imagine installing edge gutters with capacity proportional to the corresponding facet area.

Exercise 5.11. *Construct the straight skeleton polyhedron of Figure 1.8(a).*

Despite the conceptual simplicity of the straight skeleton (defined via a simple process), there seems to be no Voronoi-like definition that is not based on some procedure. Thus finding a fast algorithm for a polygon of n vertices has proved a challenge. To date, the fastest algorithm runs in (approximately) $O(n^{17/11})$ time. Although the notion of the medial axis extends naturally to 3D polyhedra (and in fact into higher dimensions), it was an unresolved issue for many years as to whether there was an unambiguous definition of the straight skeleton for a 3D polyhedron. Finally in 2008 such a definition was proposed for 3D "orthogonal polyhedra," all of whose faces meet at right angles, although extending this definition to arbitrary polyhedra seems delicate.

A surprising application of the 3D straight skeleton is to the problem of "flattening" a polyhedron — collapsing or crushing it to a plane without tearing the surface.

Exercise 5.12. *Show that the maximum number of edges of the $S(P)$ tree is $2n - 3$ for a polygon with n vertices.*

Exercise 5.13. *Design an algorithm to construct the straight skeleton in $O(n^3)$ time.*

UNSOLVED PROBLEM 20   Straight Skeleton

Find an algorithm that computes the straight skeleton of a polygon of n vertices in better than $O(n^{17/11})$ time. For example, is $O(n^{3/2})$ time achievable? Note that the size of $S(P)$ is only linear.

5.3 MINKOWSKI SUMS

Given a smooth curve C, the *offset curve* (also known as the *parallel curve*) is the locus of points offset by a constant distance r along the curve normal. That is, this new curve is offset orthogonal to the original curve C at every point. This definition relies on smoothness, which ensures a derivative at every point and thus a normal vector. An alternative definition more useful to discrete geometry is to define the offset curve as the envelope (the outer boundary of the union) of a family of disks of radius r whose centers lie on C.

Offset curves are directly related to the medial axis. An offset curve for a polygon consists of straight segments parallel to each edge and circular arcs centered on each reflex vertex, as shown in Figure 5.9(a). The traces of the joins between pairs of straight segments, and between straight segments and circular arcs, marked in part (a), form a superset of the medial axis. The traces incident to reflex vertices are not present in the medial axis. (The full set of traces is sometimes called the Voronoi diagram of the polygon.) If the circular arcs are replaced by *miter joins*,[2] we have instead the straight skeleton, as shown in Figure 5.9(b). Indeed, Figure 5.7(b) was drawn with the offset tool in ADOBE ILLUSTRATOR using the miter joins option.

Offset curves are of considerable interest in a variety of manufacturing contexts, the most prominent being *pocket machining*. A *pocket* is (usually) a shallow depression in a piece of metal cut by a cylindrical tool

---

[2] A *miter join* between two pieces of wood bevels each piece so that the junction surface bisects the angle between them.

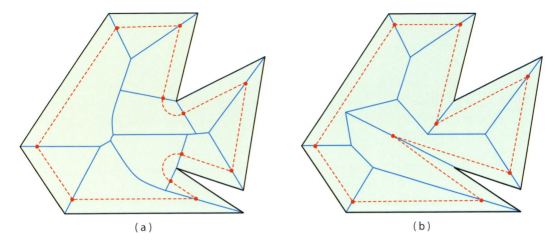

Figure 5.9. (a) Traces of the joins (in blue) between segments and arcs, which is a superset of the medial axis, and (b) the straight skeleton. The red dashes show the offset curves with circular and miter joins, respectively. Figure courtesy of Jeff Erickson.

bit. The pocket can be cut by the tool path following an offset curve of the pocket boundary, and then an offset of what remains, continuing inward until the entire pocket has been "machined" away. Another important application is defining tolerance regions for a manufactured object.

★ **Exercise 5.14.** *By analyzing the offset curve for the parabola $y = x^2$, show that an offset of a smooth curve may not itself be smooth, possibly having one or more "cusps" where the derivative is not uniquely defined.*

The definition of an offset curve is a special case of a more general concept, the Minkowski sum of two sets, to which we now turn. Let A and B be two sets of points in the plane. If we establish a coordinate system, then the points of the sets can be viewed as vectors in that coordinate system.

Definition. The *Minkowski sum* of sets A and B is

$$A \oplus B = \{x + y \mid x \in A, \ y \in B\},$$

where $x + y$ is the vector sum of the two points.

It will be a little easier to grasp the meaning of this abstract idea by considering the Minkowski sum of a single point x and a set B, defined as $x \oplus B = \{x + y \mid y \in B\}$. This is just a copy of B translated by the

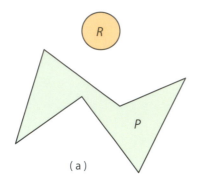

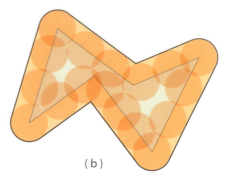

(a)   (b)

Figure 5.10. The polygon P and disk R along with their Minkowski sum on the right.

vector x, for each point y of B is moved by x. Thus $A \oplus B$ can be seen as the union of copies of B, one for each x in A.

Now suppose A is a polygon P and B is a disk R centered at the origin. Then $P \oplus R$ can be viewed as many copies of R translated by x for all x in P. Since R is centered on the origin, $x \oplus R$ will be centered on x. So $P \oplus R$ amounts to placing a copy of R centered on top of every point of P. Figure 5.10(a) shows an example of this for the polygon P and the disk R, along with their Minkowski sum in part (b). Here the Minkowski sum is the entire region on the right; we have shown P in its interior, along with selected placements of R on the boundary of P in order to clarify the addition of the objects. Indeed, $P \oplus R$ results in an "expanded" version of P, call it P^+. It should be clear that the boundary ∂P^+ of this expanded set is the outward offset of ∂P, with the normal pointing outside of the polygon.

Exercise 5.15. *What is the Minkowski sum of two squares whose sides are parallel? Describe the sum when the side lengths of the squares are a and b.*

Exercise 5.16. *Describe the Minkowski sum of a regular polygon with n vertices (of side length a) and a regular polygon with m vertices (of side length b).*

Exercise 5.17. *Prove or disprove: Any regular polygon is the Minkowski sum of a finite number of line segments in the plane.*

We now introduce one of the primary applications for the Minkowski sum in computational geometry: *motion planning*. Motion planning is the problem of designing a path for an object to move through an environment without collision with the stationary objects in that

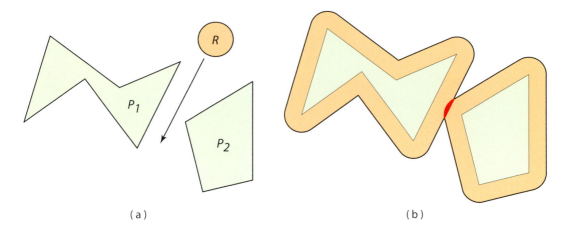

Figure 5.11. (a) Can the robot R fit through the gap between P_1 and P_2? (b) The Minkowski sums $P_1 \oplus R$ and $P_2 \oplus R$ overlap, showing that R cannot thread the gap.

environment. In a typical situation, the object is a robot and the environment has polygonal obstacles in 2D or polyhedral obstacles in 3D. Let's start with the 2D situation, where the robot is modeled as a disk R and the environment is cluttered with polygonal obstacles that must be avoided. (These assumptions are not as unrealistic as they might appear, as office mail-delivering robots are often circular.)

Consider the question whether the robot R can fit between two polygons P_1 and P_2, as illustrated in Figure 5.11(a). The robot can fit between if and only if $P_1 \oplus R$ and $P_2 \oplus R$ do not overlap, that is, if ∂P_1^+ and ∂P_2^+ do not intersect. As shown in part (b), the Minkowski sums do overlap in this example, and the robot cannot pass between the obstacles. Realize that here we are assuming R to be defined in a coordinate system whose origin is at the center of R. We will call this the *reference point* of the robot.

Having considered the case when the robot R is a disk centered at the origin, we now generalize to the situation when the robot is a convex polygon. Although rotations of R now play a significant role, we specialize to planning a *translational motion* for R. (Later, in Section 7.1, we will explore rotational motion and 3D.) Figure 5.12(a) shows the same example as Figure 5.11, but with robot R as an irregular quadrilateral. We proceed as before except that this time we need to form the Minkowski sums with $-R$ rather than R, where $-R$ is the reflection of R in the origin reference point. But why? We want ∂P^+ to be the locus of positions of the reference point when R is touching ∂P. If the point x in R touches ∂P, then x is the vector from the origin to ∂P, and so the reference point is "held away from" ∂P by the vector $-x$. This is the intuition behind the need to reflect R. The resulting Minkowski sums $P_1 \oplus -R$ and $P_2 \oplus -R$ are shown in part (b) of the figure, along with selected placements of $-R$. The boundaries ∂P_1^+ and ∂P_2^+ do not intersect, so the robot can indeed thread the gap between the obstacles.

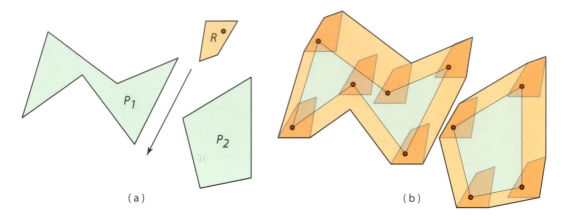

Figure 5.12. (a) Can polygon R fit through the gap between P_1 and P_2? The reference point of R is marked. (b) The Minkowski sums do not intersect, showing that R can pass through the polygons.

We have established that translational motion planning in 2D reduces to computing the Minkowski sum of each obstacle polygon P with the moving robot R. Although there are significant further issues to reach true motion planning by this approach, we set those issues aside until Section 7.1 and instead concentrate on the actual computation of $P \oplus R$.

Exercise 5.18. *Suppose the infinite plane is filled with unit square obstacles, with corners $(i - \frac{1}{2}, j - \frac{1}{2})$ and $(i + \frac{1}{2}, j + \frac{1}{2})$ for all even integers i and j. (a) If R is a disk of radius r, what is the largest value of r that permits R to move between any pair of lattice points whose coordinates are odd integers? (b) What is the largest value of r that permits R to be placed somewhere in this environment without overlapping any obstacles?*

Exercise 5.19. *Answer the questions in Exercise 5.18, now assuming the obstacles are unit disks.*

Exercise 5.20. *Show that for two convex polygons P and Q in the plane with m and n vertices, their Minkowski sum is a convex polygon with at most $m + n$ vertices.*

5.4 CONVOLUTION OF CURVES

Although we defined the Minkowski sum $A \oplus B$ as a point set, it is the *boundary curve* $\partial(A \oplus B)$ of this sum that plays the crucial role in motion planning. Constructing $A \oplus B$ and then finding its boundary is one route, but there is a method to compute the boundary more directly, via the *convolution* of the boundaries of A and B. We begin with a general definition.

Definition. Let α and β be two closed, smooth planar curves (oriented counterclockwise) whose points are interpreted as vectors in a common coordinate system. The *convolution* of curves α and β is the curve

$$\alpha * \beta = \{x + y \mid x \in A, \ y \in B, \ T_x = T_y\},$$

where T_p is the unit tangent vector at point p. We orient the curve $\alpha * \beta$ so $T_{x+y} = T_x = T_y$.

One can view $\alpha * \beta$ as constructed by rotating parallel tangents[3] around α and β and adding their contact points $x + y$. Figure 5.13 shows an example for two smooth curves along with several selected tangents. This view generalizes naturally to closed polygonal curves α and β: continuous rotation of the tangents advances the contact points in "spurts" due to the discontinuous change in the tangents. Figure 5.14 illustrates the convolution curve for $\alpha = \partial P_1$ and $\beta = \partial(-R)$ from Figure 5.12. This figure hints at the precise relationship between the $\oplus$ and $*$ operators. To elucidate this relationship, we introduce the notion of the *winding number* of a curve, an extremely useful tool in topology.

Definition. Let γ be a closed planar curve, oriented counterclockwise and possibly self-intersecting, and let x be any point of the plane not on γ. The *winding number* $\Phi_\gamma(x)$ of γ with respect to x counts the number of full revolutions of γ about x.

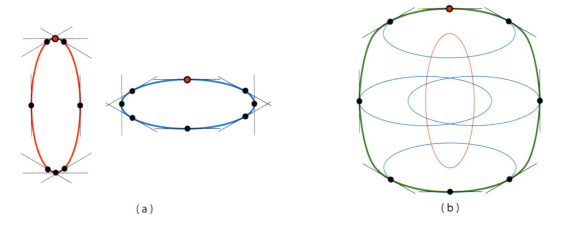

(a)   (b)

Figure 5.13. Two curves α and β and their convolution $\alpha * \beta$. The origin is at the center of both ellipses.

---

[3] These are sometimes known as *rotating calipers* because of their similarity to the parallel jaws of the measuring instrument known as a caliper.

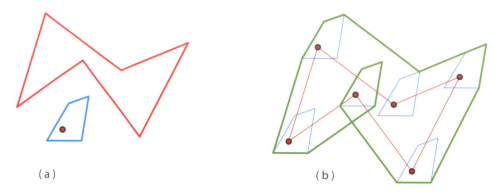

(a)   (b)

Figure 5.14. (a) Polygonal curves obtained from boundaries of P_1 and $-R$ from Figure 5.12. (b) The convolution curve $\partial P_1 * \partial(-R)$ is given in green.

Let's look at an intuitive approach to this definition: Imagine that your friend Alice walks around the curve γ (as it is oriented), starting at some point on γ and returning back to this point after a complete traversal of γ. You are standing at x and turn to face Alice at all times of her walk. The winding number is your total net angular turn during this tracking, "net" implying that negative (clockwise) turns cancel positive (counterclockwise) turns.

It should be clear that for a simple (nonintersecting) convex curve γ, the winding number $\Phi_\gamma(x) = +1$ for x inside γ and $\Phi_\gamma(x) = 0$ for x outside. Here $+1$ means one full counterclockwise turn, that is, an angle sum of 2π. Perhaps less immediate is that the same result holds true for *all* simple curves, by the Jordan curve theorem (Theorem 1.1). When the curve self-intersects, the plane is partitioned into regions with integral winding number. Two examples are given in Figure 5.15, where part (a) is a smooth curve and part (b) is the polygonal curve from Figure 5.14(b). One can see that winding number in some sense extends the Jordan curve theorem. Winding number is a key concept in algebraic topology, and generalizes to topological quantum numbers in particle physics. It is a also useful tool in many more mundane circumstances. For instance, it is used in computer graphics to detect whether a user has clicked inside and so selected an object on an interactive screen. The relationship between the Minkowski sum and convolution may now be stated as the following theorem:

Theorem 5.21. *The Minkowski sum of two planar polygons A and B is the set of points in the plane with positive winding number with respect to the convolution of ∂A with ∂B. In other words,*

$$A \oplus B = \{p \in \mathbb{R}^2 \mid \Phi_{\partial A * \partial B}(p) > 0\}.$$

Compare the Minkowski sum of the two polygons $P_1 \oplus (-R)$ in Figure 5.12(b), the convolution curve $\partial P_1 * \partial(-R)$ in Figure 5.14(b), and the winding number partition of this curve in Figure 5.15(b). The

(a)

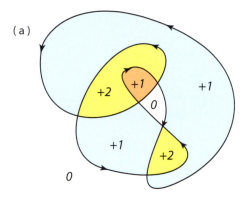

(b)

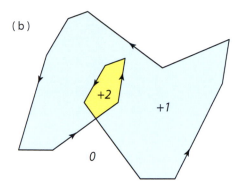

Figure 5.15. Winding numbers in regions determined by self-intersecting counter-clockwise curves.

positive winding number values of the convolution curve precisely mark the Minkowski sum.

Exercise 5.22. *What is the minimum number of vertices of a self-intersecting, closed polygonal curve that determines at least one region of the plane with winding number k?*

Exercise 5.23. *Show that if a simple, closed, counterclockwise curve γ is partitioned by a chord c into two simple, closed, counterclockwise curves $γ_1$ and $γ_2$ which share c (directed oppositely), then $Φ_γ(x) = Φ_{γ_1}(x) + Φ_{γ_2}(x)$ for any x not on γ or c.*

Exercise 5.24. *Using the previous exercise, argue that for a counterclockwise polygon P, $Φ_P(x) = 0$ for x exterior to P and $Φ_P(x) = +1$ for x interior.*

Exercise 5.25. *Use the winding number to design an O(n) time algorithm for deciding if a point x is strictly interior to a polygon.*

★ **Exercise 5.26.** *Argue that if the winding number $Φ_{∂ A*∂ B}(p)$ for a point p is positive, then p is inside $A ⊕ B$.*

Having laid the foundational relationships between the Minkowski sum, the convolution curve, and the winding number, we are now in position to focus on issues of computation. We once again assume A and B are polygons appearing in motion planning. Remember it is the boundary curve $∂ (A⊕B)$ of the Minkowski sum playing the crucial role in motion planning. Theorem 5.21 provides the machinery needed in order to compute this boundary. A breakdown of the steps is as follows:

1. Compute $\partial A * \partial B$.
2. Identify its "convolution cycles."
3. Retain cycles that have a positive winding number.
4. Merge these to construct $\partial(A \oplus B)$.

Despite the formidable intricacy of these steps, a recent implementation showed that this approach is significantly faster in practice than the best-known alternative method.

We now sketch the main idea behind just the first step, computing $\partial A * \partial B$. For notational simplicity, let $\alpha = \partial A$, $\beta = \partial B$, and $\gamma = \alpha * \beta$. Think again of parallel rotating tangents to α and β (as in Figure 5.13), touching at vertices a_i and b_j of polygons A and B, respectively. Then each rotation "event" occurs when the tangents reach the next edge on either polygon: $a_i a_{i+1}$ or $b_j b_{j+1}$, whichever is reached first. At the event, the next generated point of γ is $a_{i+1} b_j$ (if $a_i a_{i+1}$ is hit first) or $a_i b_{j+1}$ (if $b_j b_{j+1}$ is hit first).

One way to implement this is via a *star diagram* of the edge vectors, which places all edge vectors of both α and β at a common origin, labeled by their indices. Figure 5.16(a) displays two curves, α (red) and β (blue), with part (b) showing their star diagram for the 8 vectors of α and the 22 vectors of β. Fix a ray, its base at the origin of the star diagram, aiming horizontally leftward at the start. The ray represents the pair of tangents to α and β, starting at the marked topmost points of both curves. Spin this ray counterclockwise about the origin, noting when it coincides with a vector of the star diagram. As each successive edge vector is encountered, $a_{i+1} b_j$ or $a_i b_{j+1}$ is output as previously described. The spinning continues until the edges of both α and β have been circumnavigated in labeled order. Figure 5.17(a) shows the convolution curve $\alpha * \beta$ of the curves, and part (b) depicts this curve with edges colored according to their source. The coloring shows that α has already cycled once by the time the third

(a)

(b)

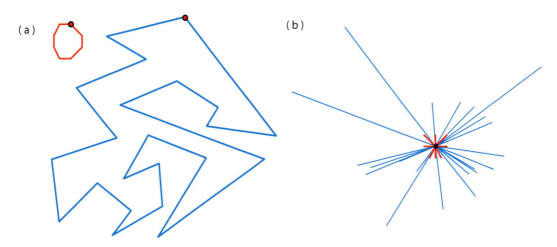

Figure 5.16. (a) Polygonal curves α and β and their (b) star diagram of edge vectors.

(a)

(b)

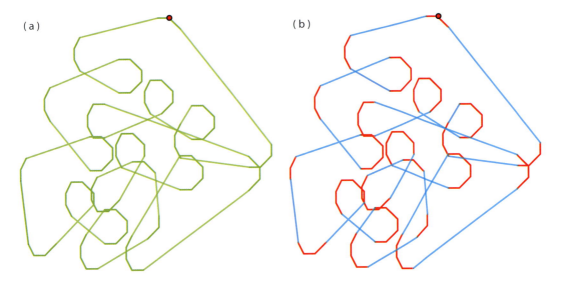

Figure 5.17. (a) The convolution curve $\alpha * \beta$ of the curves from Figure 5.16, with (b) edges colored according to their source.

edge of β is encountered, and that α cycles ten times before β cycles even once.

The many convolution cycles in Figure 5.17 indicates that the complexity of the convolution of two curves, and therefore the Minkowski sum, is potentially quadratic. In fact the complexity can be even worse, as seen in Figure 5.18. If two polygons A and B each have n vertices, the worst-case combinatorial complexity of $A \oplus B$ can be summarized by the table below. Although we have listed the complexities using the big-Oh upper-bound notation, here what is important is that in each case there are matching Ω lower bounds, that is, the bounds listed are tight.

A	B	Size of $A \oplus B$
convex	convex	$O(n)$
convex	nonconvex	$O(n^2)$
nonconvex	nonconvex	$O(n^4)$

Exercise 5.27. *Prove that the Minkowski sum of two convex polygons with n and m vertices, no pair of whose edges are parallel, has exactly $n + m$ edges.*

Exercise 5.28. *Argue that Figure 5.18 establishes that the Minkowski sum could have $\Omega(n^4)$ combinatorial complexity. Thus the bound of $O(n^4)$ time cannot be improved.*

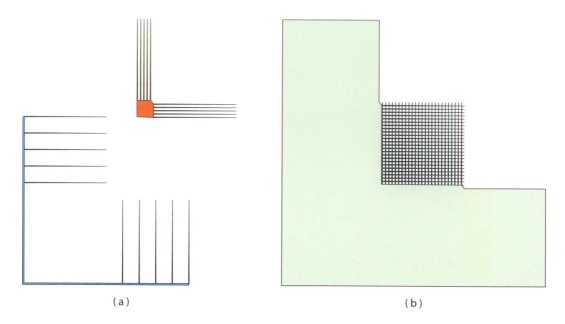

(a)   (b)

Figure 5.18. (a) Two nonconvex polygons and (b) their Minkowski sum. Figure courtesy of Ron Wein.

Exercise 5.29. *Describe simple polygons A and B, each with n vertices, whose Minkowski sum $A \oplus B$ has combinatorial complexity $O(1)$ but whose boundary convolution $\partial A * \partial B$ has combinatorial complexity $\Omega(n^4)$.*

5.5 CURVE SHORTENING

In this section we discuss a beautiful theorem that connects to the previous sections on curve offsets and morphing, and connects to polyhedra in the next chapter. The theorem is also an analog of the central technique employed in the recent resolution of the Poincaré conjecture, and so connects to the frontier of research in mathematics as well. The result is known as the "curve-shortening theorem," although it has as much to do with smoothing as it does with shortening.

We start first with the smoothing analogy. Suppose one has a jagged open curve such as that shown in Figure 5.19(a) that needs to be smoothed. A natural approach is to average nearby vertices to aggregate data and remove noise. Such smoothing is often needed to clean up noisy time-series data. A variation on this averaging idea is to perform a *midpoint transformation*: replace vertices v_i and v_{i+1} of the curve by their average, which is the midpoint of the segment $v_i v_{i+1}$. The effect of a few steps of this transformation is shown in parts (b) and (c) of the figure. Note that the curve loses one vertex per iteration — it combinatorially

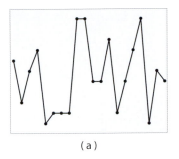 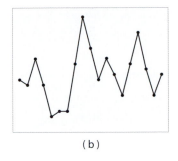 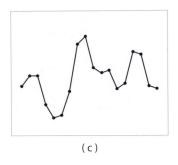

| (a) | (b) | (c) |

Figure 5.19. (a) The midpoint transformation of a 20-point open curve, which reduces (b) to 19 points after one transformation, and (c) to 18 points after two.

shortens. It also shortens geometrically, because deviations from straightness are diminished by the transformation. Applied to a closed curve, the midpoint transformation retains the original number of vertices (due to wraparound), but shortens in that the perimeter reduces, as illustrated in Figure 5.20.

One potential flaw in this technique, depending on the application, is that an initially simple polygon could become nonsimple under this transformation. We will not pursue this midpoint transformation further except to note that, in the context of curves on a surface, it is known as *Birkhoff shortening*, named after the American mathematician George Birkhoff, which plays a role in Section 6.6.

Exercise 5.30. *What is the result of applying the midpoint transformation to a regular polygon?*

Exercise 5.31. *Prove that the midpoint transformation applied to a curve (open or closed) reduces its length.*

Exercise 5.32. *Find an example where one application of the midpoint transformation changes a simple polygon to a nonsimple (self-intersecting) polygon.*

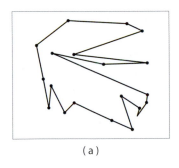 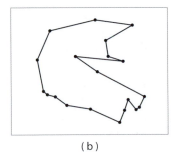 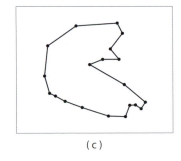

| (a) | (b) | (c) |

Figure 5.20. A 20-point closed curve retains 20 points after each iteration.

Exercise 5.33. *When the midpoint transformation is iterated, what is the limit shape? Form a conjecture.*

A second approach to smoothing/shortening a curve is to "evolve" the curve according to the heat equation, which specifies how temperature variations smooth out over time. We defer a discussion of the heat equation and start immediately with the curve shortening recipe it implies.

Let $C(s) = (x(s), y(s))$ be a smooth closed curve in the plane parametrized by arclength s. One can think of this as a unit-speed parametrization of a vehicle traveling around C. The first derivative dC/ds is the velocity of the vehicle and the second derivative d^2C/ds^2 is its acceleration. The acceleration points along the unit normal n to the curve at s, with magnitude equal to the *curvature* κ there; in other words, $d^2C/ds^2 = \kappa n$. The curvature κ is the reciprocal of the radius of the *osculating circle* at $C(s)$, the most snuggly fitting circle tangent at that point. So κ is small when the curve is nearly flat and large on sharp turns. This is the proportionality constant we feel while driving around a highway on-ramp.

Now we add a time variable t, defining a curve $C(s, t)$ for each t. We specify that the curve *evolves* with t according to the differential equation

$$\frac{\partial C}{\partial t} = \frac{\partial^2 C}{\partial s^2} = \kappa n. \tag{5.1}$$

This stipulates that each point p of the curve moves along the normal n at p with speed proportional to the curvature. This equation describes a *geometric flow*, an evolution of the geometry of C over time t.

Example 5.34. To clarify the meaning of equation (5.1), we explicitly solve it for the simplest possible closed curve, a circle. Circles are easily parametrized by arc length as $C(s) = (\cos s, \sin s)$ for s in $[0, 2\pi]$. Let's guess that $C(s, t)$ can be represented as $f(t)(\cos s, \sin s)$, reasonable given the symmetry of the circle. Then equation (5.1) becomes

$$\frac{\partial C}{\partial t} = f'(t)(\cos s, \sin s) = f(t)(-\cos s, -\sin s),$$

where $f'(t) = \partial f/\partial t$. Solving $f'(t) = -f(t)$ leads to $f(t) = e^{-t}$. So the evolving family of curves is

$$C(s, t) = e^{-t}(\cos s, \sin s).$$

At each time $t > 0$, this is a scaled, concentric version of the original circle, reduced by the factor e^{-t}, which approaches 0 as t goes to infinity.

We can now state the curve-shortening theorem.

Theorem 5.35 (Curve Shortening). *Every smooth, simple closed curve C evolves under the flow defined by equation (5.1) so that it remains simple for all time and converges to a round point.*

Here converging to a *round point* means converging to a circle whose radius goes to zero as t approaches infinity. Clearly this implies both smoothing and shortening. Perhaps what is most remarkable about this theorem is the avoidance of self-intersections over all time. This means that a spiral like that in Figure 5.21 will evolve without crossing itself — it somewhat magically uncurls! In fact, the simultaneous evolutions of a pair of nested curves never bump into each other: the inner one outruns the collapsing outer one!

This remarkable theorem was first proved in the 1980s by Michael Gage and Richard Hamilton for convex curves, and then extended by Matthew Grayson to nonconvex curves and curves on surfaces. It is now known as the Gage-Hamilton-Grayson theorem. We will reencounter Richard Hamilton when we discuss the Poincaré conjecture at the end of Section 5.6.

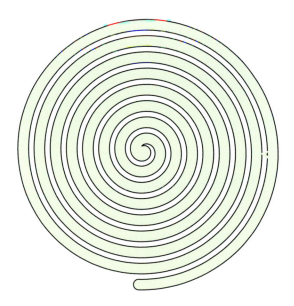

Figure 5.21. Continuous curve shortening convexifies this curve without self-intersection.

We now consider a discretization of the methods above, looking at curve shortening of polygons. The curve-shortening flow in equation (5.1) relies on derivatives and so does not apply to polygons which have no unique tangent or normal vectors at their vertices. Various discrete flows analogous to equation (5.1) have been explored. Here we describe an especially simple one suggested by Bennett Chow and David Glickenstein in 2007.

In this flow, the continuous morphing a curve becomes a discrete replacement of an n-sided polygon P by a new n-sided polygon P', just as in the midpoint transformation. Figure 5.22 shows that the term

$$n_i = (v_{i+1} - v_i) + (v_{i-1} - v_i) \tag{5.2}$$

crudely approximates an inward-pointing normal vector at v_i. We now mimic equation (5.1) by moving each vertex v_i of P to a new location given by

$$v_i' = v_i + \delta n_i,$$

where $\delta > 0$ is a small step-size scale factor. As δ goes to 0, $(v_i' - v_i)/\delta$ approaches dv_i/dt, and so the discrete flow equation may be written as

$$\frac{dv_i}{dt} = n_i. \tag{5.3}$$

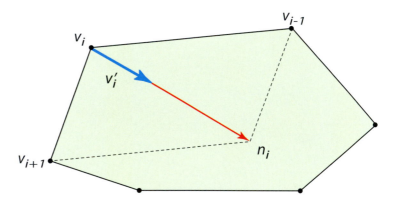

Figure 5.22. Construction of the vectors n_i and v_i'.

This describes the continuous evolution of each discrete vertex of P. The resemblance with equation (5.1) will be further elucidated after we explore the discrete flow. Figure 5.23 illustrates the flow on a 20-vertex polygon. Note it has the same characteristics claimed for the smooth case. Indeed, Chow and Glickenstein prove the following:

Theorem 5.36 (Discrete Curve Shortening). *Every simple polygon evolves under the flow defined by equation (5.3) so that it converges* *to a point whose shape is asymptotically an affine transformation of a regular polygon.*

Here an *affine transformation* is a linear distortion that maintains parallel lines, so an affine regular polygon is a distorted version of a regular polygon, the analog of the "round point" in Theorem 5.35. Theorem 5.36 is a faithful analog of Theorem 5.35 except that there is no guarantee that if the initial polygon is simple, all intermediate shapes will also be simple. Nevertheless, with a dense sampling of vertices, the discrete flow generally avoids self-intersection, as shown in Figure 5.24.

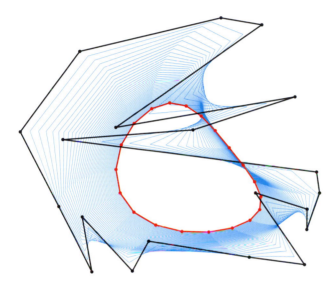

Figure 5.23. A discrete flow of a simple 20-gon (in black) with 40 iterations using $\delta = 1/10$.

UNSOLVED PROBLEM 21   **Discrete Flow**

Define an analogous discrete polygon flow that guarantees simplicity for all time, for sufficiently small time steps δ. The rules for movement of each vertex v_i should depend only on the local neighborhood of v_i along ∂P. Perhaps it will be necessary to restrict to a subclass of all polygons.

Curve shortening is used in a variety of circumstances where smoothing is needed, such as smoothing shapes in images. To obtain smoothing without shortening, the arclength of the curve is renormalized at each infinitesimal step (as in the lower sequence of Figure 5.24). A rather surprising application of discrete curve shortening is the *rendezvous problem* for mobile autonomous robots: how each robot should behave in

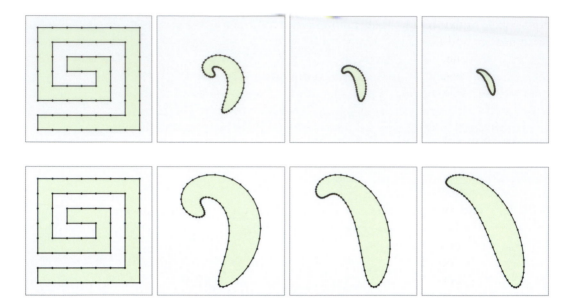

Figure 5.24. The discrete flow is applied to a spiral 64-gon with 100 iterations with $\delta = 1/5$ between frames. The top row is to scale whereas the bottom row shows the same sequence rescaled.

order to meet at a common point. The traces of the vertices in Figure 5.23 can be interpreted as paths of twenty robots converging toward a meeting point. Yet another application is to find closed geodesics on surfaces, which we will explore in Section 6.6.

Exercise 5.37. *What is the result of applying the discrete flow of equation (5.3) to a regular polygon?*

Exercise 5.38. *Find an example where the discrete flow on an initially simple polygon passes through a nonsimple polygon.*

Exercise 5.39. *In terms of the original shape, what is the point to which the flow converges? Form a conjecture.*

Exercise 5.40. *Suppose the discrete flow is applied to a 3D polygon, a nonplanar closed chain of segments in $\mathbb{R}^3$. Formulate a conjecture on what happens.*

5.6 THE HEAT EQUATION

We now indicate how both the smooth flow of equation (5.1) and the discrete flow of equation (5.3) can be seen as geometric analogs of a more general flow, described by the *heat equation*. The heat equation is arguably the most important partial differential equation in mathematics and physics. It describes the distribution of heat in a given region over time, related to the study of Brownian motion, chemical diffusion, and several other related processes. Let $u(x, t)$ be a function that expresses the heat at a point x in a uniform medium at time t. The heat equation

$$\frac{\partial u}{\partial t} = \frac{\partial^2 u}{\partial x^2} \tag{5.4}$$

describes the time evolution of the temperature at each point x.

Since heat tends to even out over time, the equation describes a smoothing and averaging process: The heat $u(x)$ at a particular spot x is affected by the heat at nearby spots $u(x - \delta)$ and $u(x + \delta)$. If, for example, $u(x - \delta)$ is colder than $u(x)$ and $u(x + \delta)$ hotter by the same amount, then the tugs cancel out their effects at $u(x)$. But if the average surrounding temperature $\frac{1}{2}[u(x-\delta)+u(x+\delta)]$ differs from $u(x)$, it pulls the temperature at $u(x)$ with a tug proportional to the difference,

$$u(x) - \frac{1}{2}[u(x - \delta) + u(x + \delta)] = \frac{1}{2}\Big([u(x) - u(x - \delta)] - [u(x + \delta) - u(x)]\Big).$$

This expression is easily recognizable as the second derivative with respect to x: a difference of differences.

We will not attempt precise derivations but content ourselves with pointing out a formal symbolic similarity between the curve-shortening equations and the heat equation. Replacing x in equation (5.4) with s and $u(x, t)$ with $C(s, t)$ yields equation (5.1) directly. Turning to equation (5.3), we can think of the derivative at the midpoint v_i^* of $v_i v_{i+1}$ as approximated by

$$\frac{\partial v_i^*}{\partial s} \approx v_{i+1} - v_i.$$

Accepting this, we can approximate the second derivative:

$$\frac{\partial^2 v_i}{\partial s^2} \approx \frac{\partial v_i^* - \partial v_{i-1}^*}{\partial s} \approx (v_{i+1} - v_i) - (v_i - v_{i-1}).$$

As the right-hand side is precisely n_i from equation (5.2), we may view equation (5.3) as

$$\frac{\partial v_i}{\partial t} = \frac{\partial^2 v_i}{\partial s^2}.$$

Although we are performing only symbol manipulations here, and employing approximations, the affinity of the continuous curve-shortening flow, the discrete flow, and the heat diffusion equation should now be evident.

Yet one more analogy brings us to the Poincaré conjecture. It would take us very far afield to explain this complex and advanced topic adequately. Instead we attempt to sketch enough of the story so that the relationship to curve shortening can be discerned. Henri Poincaré, a brilliant French mathematician and physicist, and one of the founders of the field of topology, formulated the following conjecture around 1900.

Poincaré Conjecture. *Every simply connected closed 3-manifold is homeomorphic to the 3-sphere.*

We now describe the terms used in this mysterious statement.

1. A *manifold* is a space that is locally Euclidean, in that it looks like Euclidean space in the neighborhood of each point. A *3-manifold* is one that is locally like $\mathbb{R}^3$, our familiar 3D space. But globally it might have a different structure.
2. A *3-sphere* is the analog of the standard sphere (a 2-sphere) in one higher dimension. It can be viewed as the set of all points in $\mathbb{R}^4$ that are a fixed distance from the origin, the center of the 3-sphere. A 3-sphere is a 3-manifold.
3. A *closed* 3-manifold is finitely bounded, unlike $\mathbb{R}^3$, which is unbounded. A circle, a sphere, and a 3-sphere are all closed manifolds. Our universe is some type of 3-manifold, but whether closed or unbounded is unknown.
4. Two manifolds are *homeomorphic* if there is a continuous bijection (called a homeomorphism) from one to the other, whose inverse is also continuous. Roughly, this means that one manifold can be continuously deformed into the other by stretching and bending.
5. A manifold is *simply connected* if every closed curve (*loop*) can be contracted to a point while staying in the manifold. A *torus* (the surface of a donut) is not simply connected, whereas a sphere is.

The two-dimensional version of the conjecture was already established by 1900: if every loop can be contracted in a particular closed 2-manifold, then that manifold must be homeomorphic to the 2-sphere. Poincaré conjectured that the same holds for 3-manifolds: if a 3-manifold is simply connected, it must be homeomorphic to the 3-sphere. And the same question may be asked for n-dimensional manifolds. It took 60 years before any progress was made when, in 1961, Stephen Smale shocked the mathematics community by proving the conjecture for dimensions $n \geq 5$. For this, he earned the Fields Medal, the highest honor a mathematician can receive for his or her research. Twenty years later, in 1982, Michael Freedman proved the Poincaré conjecture for dimension $n = 4$ using vastly different techniques. He was also awarded the Fields Medal for his work.

The last remaining and most interesting case (because we live in a 3-manifold) was finally settled almost exactly a century after Poincaré formulated it: Grigori Perelman proved in 2003 that the Poincaré conjecture indeed holds true in 3D. As mathematicians have come to expect, Perelman was awarded the Fields Medal for his work in 2006. In a stunning move, however, Perelman refused the honor, saying that if his proof is correct then no other recognition is needed. In 2010, Perelman was also awarded the first Clay Millennium Prize award of one million dollars because settling the Poincaré conjecture is one of the seven "Millennium problems." Consistent with his principles, he declined this prize as well.

A key to Perelman's resolution of the conjecture is the *Ricci flow* introduced by Richard Hamilton in the early 1980s. The *Ricci tensor* **Ric** is a multidimensional generalization of the curvature κ we encountered in equation (5.1). Rather than defining a curve deformation flow as in that equation, the Ricci flow defines a *metric* deformation flow, where a *metric tensor* g specifies how lengths and angles vary at different points of the manifold. In a sense, g determines the local shape of the manifold. Hamilton's flow equation is written as

$$\frac{\partial g}{\partial t} = -2\,\mathbf{Ric}(g). \qquad (5.5)$$

Very roughly, this specifies that the manifold should "shrink" around every point at a rate proportional to the Ricci curvature there (which, when negative, implies expansion).[4] If one views variations in the metric g as describing a "lumpy" manifold, then, just as in curve shortening, the Ricci flow should tend to lower the mountains, raise the valleys, and in general smooth out local irregularities in the metric. If this could be shown to smooth any 3-manifold metric into the homogeneous metric of a 3-sphere, then the Poincaré conjecture would be settled.

It was in this context that the Gage-Hamilton-Grayson theorem was discovered. But unlike the elegant non-self-intersection property established in that theorem, Ricci flow can lead to extreme "neck-pinch singularities," which presented a serious stumbling block to completing Hamilton's proof plan. Perelman's breakthrough was to see how to classify and tame the singularities in the Ricci flow, ultimately leading to the proof that simply connected 3-manifolds are homeomorphic to the 3-sphere.

---

[4] Although the factor 2 is just a convention, the minus sign fixes directionality, ensuring solutions for positive time.

Returning to equation (5.5), we continue the symbolic analogy between the various equations encountered in this chapter. Crudely, the Ricci tensor measures the volume distortion of a small ball centered at a point on the manifold in comparison to its volume in flat Euclidean space. Like its analog κ, it is essentially a second derivative. Reducing the Ricci tensor to a 1D variable x (where, admittedly, it loses much of its sense), it may be expanded in a Taylor series as

$$\mathbf{Ric}(g) = \frac{\partial^2 g}{\partial x^2} + \cdots,$$

where the omitted terms are of higher order.[5] Thus the Ricci flow equation can be viewed as yet another analog of the heat equation. The four analogous equations are displayed in the table below to emphasize their structural similarity. Curve shortening smooths out sharp curvature

Smooth Curve Shortening	Discrete Curve Shortening	Heat Equation	Ricci Flow
$\dfrac{\partial C}{\partial t} = \dfrac{\partial^2 C}{\partial s^2}$	$\dfrac{\partial v_i}{\partial t} = \dfrac{\partial^2 v_i}{\partial x^2}$	$\dfrac{\partial u}{\partial t} = \dfrac{\partial^2 u}{\partial x^2}$	$\dfrac{\partial g}{\partial t} \approx -2\dfrac{\partial^2 g}{\partial x^2}$

by moving normal to the curve at a rate proportional to curvature, whereas discrete curve shortening approximates the same process in a discrete setting. The heat equation describes the diffusion of heat from hot spots "downhill" to cool spots, and the Ricci flow smooths manifolds by contracting regions of positive curvature and expanding regions of negative curvature. It is a remarkable path from the obvious idea of averaging to smooth time-series data to the resolution of the Poincaré conjecture!

5.7 CURVE RECONSTRUCTION

There are many scanning devices today that rapidly and accurately collect a dense sample of points from the surface of a 3D object, such as laser range finders, stereoscopic photography, and laser scanners. The raw data collected by these devices are generally represented by 3D coordinates of points from the surface. *Reconstruction* of the surface is the task of developing a representation of the surface by connecting nearby points into some type of *mesh*, a surface of triangles. This difficult and important

---

[5] In 2D, the differential term takes the more familiar form $(\frac{\partial^2}{\partial x^2} + \frac{\partial^2}{\partial y^2})g$, often written as Δg or $\nabla^2 g$.

problem is the focus of intense activity today. The examples in Figure 5.25 demonstrate how far the technology has advanced in this area.

A simpler version of the problem is *curve reconstruction*: Given a set of point coordinates sampled from some curve C, connect those points adjacent along C to form a polygonal curve P that represents C. After early heuristics for this "connect-the-dots" problem, a breakthrough paper by Nina Amenta, Marshall Bern, and David Eppstein in 1998 specified the problem in a way that permitted provably correct algorithms. We will present their CRUST algorithm, which relies heavily on the properties of the medial axis, the Voronoi diagram, and the Delaunay triangulation.

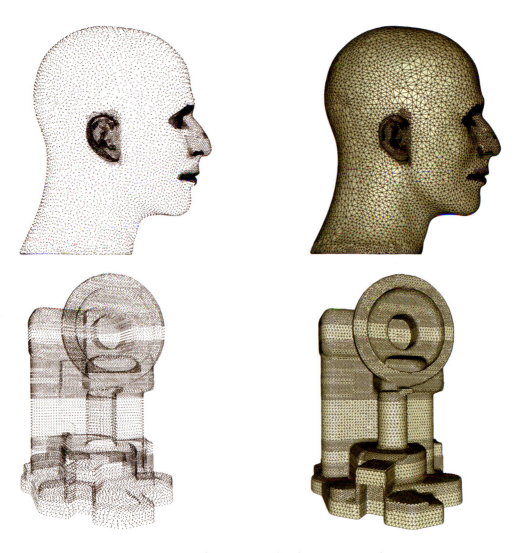

Figure 5.25. Surface reconstructions from point clouds in 3D via the TIGHT COCONE algorithm. The top row is a mannequin with 16216 vertices and 32308 triangles and the bottom row is an oil pump with 30927 vertices and 61850 triangles. Figures courtesy of Tamal Dey.

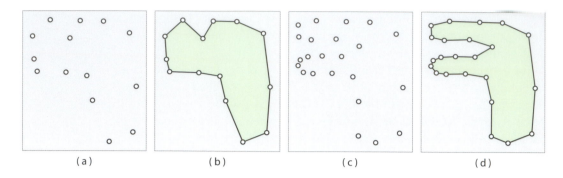

Figure 5.26. Parts (a) and (c) show two samples from the same curve, the first a subset of the second. The reconstructions (b) and (d) show that (a) was under-sampled.

This algorithm has served as the inspiration of nearly every subsequent algorithm developed in the last decade, both for curve and for surface reconstruction.

It is obvious that accurate reconstruction requires a dense enough sample. Figure 5.26(a) and (c) show two samples from the same curve, the first a subset of the second. The reconstructions of the curve from these samples in parts (b) and (d), respectively, show that (a) is not a dense enough sample. It is also clear that what is "dense enough" varies with the "complexity" of portions of the curve. One of the key advances was to define a precise notion of "dense enough" via the medial axis, a notion called the *local feature size*.

Definition. Let C be a smooth closed curve in the plane, and let x be a point of C. The *local feature size* $\rho(x)$ of x is the shortest distance from x to the medial axis of C.

In Section 5.1 we defined the medial axis $M(C)$ of a closed smooth curve C as the locus of the centers of disks that touch C at two or more distinct points. Although we emphasized the portion of the medial inside C, for a nonconvex curve, $M(C)$ has branches both interior and exterior to C, and both are relevant for the definition of local feature size. Figure 5.27(a) shows a closed curve along with its (approximate) medial axis, located both within and outside the curve. Part (b) of the figure shows a few of the disks whose centers define the medial axis. Note that $\rho(x)$ is small for points on high-curvature sections of C and potentially larger on low-curvature sections.

Exercise 5.41. *Show that the hedge in the preceding sentence —* "potentially larger" *— is necessary by constructing a curve C with a point x in C such that $\rho(x)$ is arbitrarily small, even if the curvature of C in a neighborhood of x is zero.*

(a)

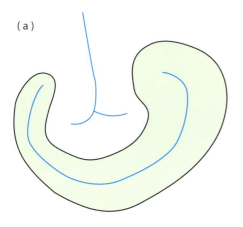

(b)

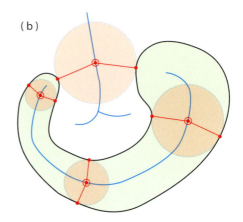

Figure 5.27. (a) A closed curve C along with its (approximate) medial axis $M(C)$. (b) A few of the disks whose centers define the medial axis and determine the local feature size.

Now we can define the proper notion of density for our sampling:

Definition. Let $0 < \varepsilon < 1$. A set S of points sampled from C is an *ε-sample* if each point x in C has a point p in the sample S, where $|x - p| \leq \varepsilon \rho(x)$.

Note that this definition forces the sample to be dense in sections of C that are "complicated" in that $M(C)$ is near by. One more definition will enable us to state the CRUST algorithm.

Definition. A *correct polygon reconstruction* P of a curve C from a sample S connects points p and q in P if and only if p and q are consecutive sample points along C.

The points p and q are consecutive if the curve segment between p and q is empty of other points of S. Figure 5.28(a) shows a sample set of the curve in Figure 5.27 and (b) gives a correct polygon reconstruction of the curve from this sample. The goal of curve reconstruction is to find an algorithm that guarantees correct polygonal reconstruction from an ε-sample, for some particular $\varepsilon > 0$. The CRUST algorithm, to which we now turn, achieves this provable correctness for $\varepsilon < 1/5$.

Recall from Section 3.4 that an edge e is in the Delaunay triangulation $\mathrm{Del}(S)$ of a set of points S if and only if e has a circumscribing disk empty of other points of S. Consequently, if the sample points are dense enough, then edge e of the correct polygonal reconstruction is an edge of $\mathrm{Del}(S)$. Thus the Delaunay triangulation contains the edges we seek, and the remaining challenge is to avoid the incorrect edges of $\mathrm{Del}(S)$. Figure 5.29(a) shows $\mathrm{Del}(S)$ of the sample set S from Figure 5.28(a). Notice that the edges needed for proper reconstruction of S, as given in

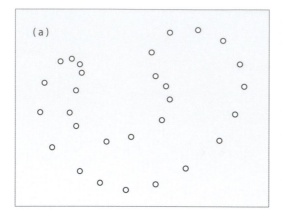

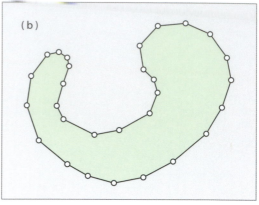

Figure 5.28. (a) A sample set of the curve in Figure 5.27, along with (b) a correct polygon reconstruction of the curve from the sample.

Figure 5.28(b), are a subset of Del(S). The CRUST algorithm is designed for the very purpose of finding the correct edges from the set of Delaunay edges.

Several key insights lead to the CRUST algorithm. We list them first on an intuitive level, for sufficiently small ε, and illustrate these claims through Figure 5.29.

1. The Voronoi vertices V of Vor(S) lie near $M(C)$.
2. Any circumscribing disk of an incorrect edge of Del(S) crosses the medial axis $M(C)$ of C.
3. An incorrect edge e of Del(S) cannot also appear in Del($S \cup V$) because a circumscribing disk for e contains a vertex in V.
4. Each correct edge of Del(S) also appears in Del($S \cup V$).

Claim 1 can be seen by comparing the black Voronoi vertices V in Figure 5.29(b) to the medial axis in Figure 5.27. Of course, a Voronoi vertex has the property that the Voronoi disk centered there touches at least three points of S, so it is natural that the points of V fall near $M(C)$. Claim 2 is especially evident with the internal diagonals of Del(S) which clearly cross the medial axis $M(C)$. It then makes sense (Claim 3) that a circumscribing disk of such an incorrect edge must include a vertex in V, because V approximates $M(C)$.

Finally, Claim 4 can be seen by centering a circumscribing disk for a correct edge $e = ab$ on the point x of C crossed by the perpendicular bisector of ab. Figure 5.30 provides an illustration. The distance from x to the medial axis is the local feature size $\rho(x)$. But because S is an ε-sample, $|a - x| \leq \varepsilon\, \rho(x)$. Thus the radius $|a - x|$ of this circumscribing disk will not reach $M(C)$ for $\varepsilon < 1$, and so will not contain a point of V for sufficiently smaller ε. This empty disk guarantees that e is part of Del($S \cup V$). This intuitive justification leads to the remarkably simple CRUST algorithm.

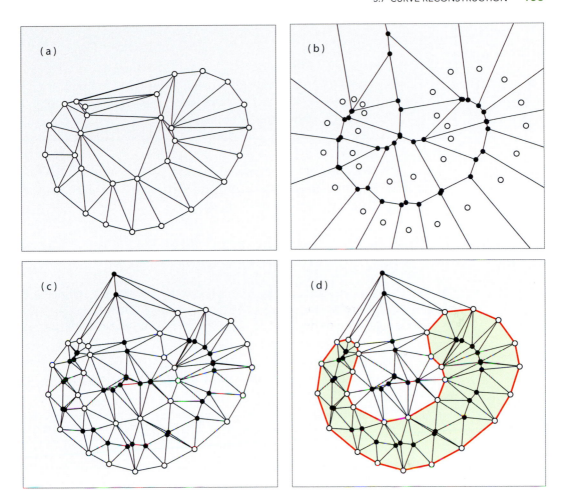

Figure 5.29. (a) The Delaunay triangulation and (b) the Voronoi diagram of the sample S from Figure 5.28(a). The Voronoi vertices V are shown in black. (c) The Delaunay triangulation of $S \cup V$ and (d) its edges having both endpoints in S, marked in red.

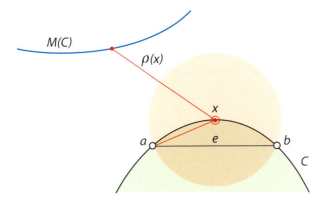

Figure 5.30. A circumscribing disk for a correct edge e will not reach $M(C)$.

> **CRUST**   Curve Reconstruction Algorithm
>
> Let S be the set of sample points. Compute the Voronoi diagram $Vor(S)$ of S and let V be its set of Voronoi vertices. Compute the Delaunay triangulation $Del(S \cup V)$. The curve P is composed of the edges of $Del(S \cup V)$ with both endpoints in S.

The "provable correctness" of this algorithm may be stated in the following theorem. Proving this theorem formally requires careful analysis to justify the claims we left intuitive above.

Theorem 5.42. *The* CRUST *algorithm outputs the correct polygonal reconstruction whenever S is an ε-sample with ε < 1/5.*

One goal of subsequent research has been to increase ε while still maintaining provable correctness. This was achieved for $\varepsilon < 1/3$ by Tamal Dey using the NN-CRUST algorithm (NN stands for "nearest neighbor"), which has the added advantage of only computing one Delaunay triangulation rather than the two computations in the CRUST algorithm. Subsequent improvements have reached $\varepsilon < 1/2$. Finally, as mentioned earlier, the CRUST algorithm and several of its descendants have been generalized for surface reconstruction, an area of active research.

> **UNSOLVED PROBLEM 22**   Curve Reconstruction
>
> Find an algorithm that guarantees correct curve reconstruction from an ε-sample for some $\varepsilon \geq 1/2$.

Exercise 5.43. *Prove that any disk centered at a point x in C with radius less than or equal to $\rho(x)$ intersects C in a connected subcurve of C.*

Exercise 5.44. *Prove that if a disk is tangent to C at a smooth point x in C and has radius less than or equal to $\rho(x)$, then the disk contains no points of C in its interior.*

Exercise 5.45. *Prove that if x and y are points in C, then $\rho(x) \leq \rho(y) + |x - y|$. This property is known as Lipschitz continuity.*

SUGGESTED READINGS

Francis Chin, Jack Snoeyink, and Cao An Wang. Finding the medial axis of a simple polygon in linear time. *Discrete and Computational Geometry*, Volume 21, pages 405–420, 1999.
> The linear-time algorithm for computing the medial axis of a polygon successively partitions the polygon into three different varieties of "histograms."

Jeff Erickson and David Eppstein. Raising roofs, crashing cycles, and playing pool: Applications of a data structure for finding pairwise interactions. *Discrete and Computational Geometry*, Volume 22, pages 569–592, 1999.
> This paper describes the fastest known algorithm for computing the straight skeleton, as well as a good introduction to the topic. The applications of the straight skeleton to origami and flattening are described in *Geometric Folding Algorithms: Linkages, Origami, Polyhedra* (Erik D. Demaine and Joseph O'Rourke, Cambridge University Press, 2007).

Ron Wein. 2D Minkowski sums, Chapter 22 of *The CGAL Manual*. http://www.cgal.org/Manual/.
> A clear description of computing Minkowski sums via convolutions.

Kai-Seng Chou and Xi-Ping Zhu. *The Curve Shortening Problem*. Chapman & Hall, 2001.
> A rigorous and technically demanding book-length treatment of the mathematical aspects of curve shortening.

John Morgan and Gang Tian. *Ricci Flow and the Poincaré conjecture*, *Clay Mathematics Monographs*, Volume 3. American Mathematical Society, 2007.
> Full details of the complete proof of the Poincaré conjecture in 500 pages, at the advanced graduate level.

Tamal Dey. *Curve and Surface Reconstruction: Algorithms with Mathematical Analysis*. Cambridge University Press, 2006.
> This demanding but clear book by one of the leaders of the field covers curve and surface reconstruction thoroughly, including the CRUST algorithm.

6 POLYHEDRA

We have already encountered polyhedra several times in this book. In this chapter, we study them more systematically with the dual goal of strengthening 3D intuition and presenting several theorem gems. We start with the Platonic solids (Section 6.1) and then revisit Euler's formula (Theorem 3.12) in the polyhedral context (Section 6.2) in which it originated. We follow that with two beautiful and useful theorems: The Gauss-Bonnet theorem (Section 6.3) and Cauchy's rigidity theorem (Section 6.4). We then study shortest paths on convex polyhedra (Section 6.5), a topic which brings us back to the ubiquitous Voronoi diagram. We close the chapter with a deep theorem of Lyusternik and Schnirelmann on closed geodesics on polyhedra (Section 6.6), which connects back to our discussion of curve shortening from Section 5.5.

6.1 PLATONIC SOLIDS

A polyhedron is the natural generalization of a two-dimensional polygon to three dimensions: it is a bounded region of space whose boundary is composed of a finite number of flat polygonal *faces*, any pair of which either are disjoint or meet at edges and vertices. Before making this definition more precise (a surprisingly delicate task), we begin with convex polyhedra, and in particular, the most famous polyhedra, the Platonic solids.

Convex polyhedra are the natural generalizations of convex polygons to 3D. A *convex polyhedron* P satisfies the definition of convexity in that the segment xy connecting any two points x and y of P is contained inside P. Just as convex polygons can be characterized by the local boundary condition that each vertex be convex, convex polyhedra can be specified locally via *dihedral angles*, as discussed in Section 1.5. An edge whose dihedral angle is at most π is called a *convex edge*; otherwise, a *reflex edge*. Then the local boundary condition for convexity can be stated as follows:

Lemma 6.1. *A polyhedron is convex if and only if all of its edges are convex.*

(a)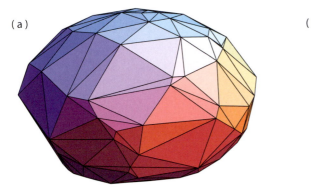

(b)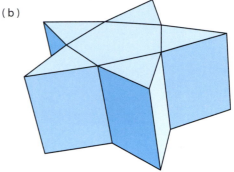

Figure 6.1. (a) A convex polyhedron and (b) a nonconvex polyhedron, a type of pentagrammic prism.

We do not prove this now, as it is best understood in terms of spherical polygons introduced later (see ahead to Figure 6.20 and Exercise 6.41). Figure 6.1(a) shows a typical convex polyhedron, and (b) shows a particular nonconvex polyhedron with five reflex edges. This latter figure shows that having all convex faces is not sufficient to ensure that the polyhedron is convex. However, it is a necessary condition, as claimed by the following exercise.

Exercise 6.2. *Show that each face of a convex polyhedron must be a convex polygon.*

Another important source of boundary information is provided by the *face angles* of the polyhedron, the angles around each polygonal face of P, and the sum of these face angles incident to each vertex. We will need this consequence of convexity at several points in the chapter:

Lemma 6.3. *For any convex polyhedron, the sum of the face angles incident to each vertex is at most 2π.*

Again a proof is easiest in terms of spherical polygons, and we again defer (Exercise 6.42). It is natural to assume that this necessary consequence of convexity is also sufficient, but there are nonconvex polyhedra with the face angle sum at each vertex at most 2π. A somewhat subtle example is provided by Figure 6.3(b), where 10 face angles each of $\frac{1}{5}\pi$ meet at each vertex. But a simpler example can be attained by "denting" a convex polyhedron, such as that shown later in Figure 6.16.

Now we turn to the study of *regular* convex polyhedra. A regular polygon is one with equal side lengths and equal angles, such as the equilateral triangle, the square, and so on. Clearly there are an infinite

Figure 6.2. The Platonic solids: tetrahedron, cube, octahedron, dodecahedron, icosahedron.

variety of regular polygons with n sides, one for each $n > 2$. It is then natural to examine *regular polyhedra*, although the appropriate definition of regularity is no longer so evident.

Definition. A convex polyhedron is *regular* if all its faces are congruent regular polygons, and the number of faces incident to each vertex is the same for all vertices.

Notice that there is no mention of dihedral angles in this definition. It turns out that these condition imply equal dihedral angles, so that fact need not be included in the definition. Unlike the situation in 2D, the surprising implication of these regularity conditions is that there are only a finite number of distinct regular polyhedra in 3D, the five *Platonic solids*. These polyhedra, named the *tetrahedron*, the *cube*, the *octahedron*, the *dodecahedron*, and the *icosahedron*, are shown in Figure 6.2. They are named the *Platonic* solids because they are discussed in Plato's *Timaeus*.

We now prove that there are exactly five regular polyhedra, under the strict definition of regularity above. This theorem was given by Euclid (*Elements*, Book XIII) as the culminating proposition:

> "I say next that no other figure, besides the said five figures, can be constructed which is contained by equilateral and equiangular figures equal to one another."

It was proved without any algebra, which of course had not yet been invented. The intuition behind the proof we present is that the internal angles of a regular polygon grow large with the number of vertices of the polygon, but there is only so much room to pack these angles around each vertex of a regular polyhedron while still satisfying Lemma 6.3.

Theorem 6.4. *The five Platonic solids are the only regular polyhedra.*

Proof. Assume we are given a regular polyhedron P. Let k be the number of vertices per face of P. By Exercise 1.11, the sum of the interior angles of a k-gon is $\pi(k-2)$. And since the faces of P are regular, each face angle must be $\pi(1-2/k)$.

Let's insist that a vertex be a true corner of the polyhedron, one that you could feel with your fingers were you to touch it there. Then the sum of the face angles incident to a vertex is strictly less than 2π. Let m be the number of faces incident to each vertex, which is the same for every vertex by our definition of regularity. Then we have m face angles, each $\pi(1 - 2/k)$, which must sum to less than 2π. A bit of algebraic manipulation of this constraint leads to a particularly revealing form:

$$m\pi(1 - 2/k) < 2\pi. \tag{6.1}$$

$$1 - 2/k < 2/m.$$

$$km < 2m + 2k.$$

$$km - 2m - 2k + 4 < 4.$$

$$(k - 2)(m - 2) < 4. \tag{6.2}$$

We know that both k and m are integers. Because each face must be at least a triangle, $k \geq 3$. Moreover, $m \geq 3$ since at least three faces must meet at each vertex; otherwise the polyhedron would collapse to planarity there. This leaves only the five $\{k, m\}$ combinations listed in the table below, resulting in the five Platonic solids. $\square$

Name	k-gon	m per vertex	$(k-2)(m-2)$	V	E	F
Tetrahedron	3	3	1	4	6	4
Cube	4	3	2	8	12	6
Octahedron	3	4	2	6	12	8
Dodecahedron	5	3	3	20	30	12
Icosahedron	3	5	3	12	30	20

Several remarks on the proof are in order. The strict inequality $< 2\pi$ in equation (6.1) leads to the strict inequality < 4 in equation (6.2). Just barely violating this latter inequality, with $k = m = 4$ when $(k - 2)(m - 2) = 4$, leads to gluing four squares at every vertex. This renders the vertex flat and fails to achieve a closed polyhedron. The same can be said for $k = 6$ and $m = 3$, or $k = 3$ and $m = 6$, when again $(k - 2)(m - 2) = 4$. The former corresponds to gluing three regular hexagons at each vertex, the latter to gluing six equilateral triangles at each vertex. In either case, a flat 2π vertex results. The constraint $m \geq 3$ implicitly rules out a flat, doubly covered regular k-gon as a regular polyhedron. Indeed, the official definition of a polyhedron below will exclude this, but there are circumstances where doubly covered convex polygons are naturally viewed as degenerate polyhedra.

That the listed local information determined by k and m leads to the polyhedra whose global properties are described in the table is not immediately obvious, but it is easy to check that the polyhedra indeed achieve the $\{k, m\}$ values. This pair of numbers $\{k, m\}$ is called the *Schläfli symbol*, in honor of the nineteenth-century Swiss mathematician Ludwig Schläfli, who developed a notational system to record the structure of regular and semi-regular polyhedra.

Loosening the definition of regularity to permit several different regular polygons as faces, but still meeting in the same vertex configurations, leads to the 13 Archimedean solids, including the truncated icosahedron (soccer ball) composed of 12 pentagons and 20 hexagons, as shown in Figure 6.3(a). Loosening further to permit nonconvexity leads to the 75 *uniform polyhedra*, including, for example, the great dodecahedron shown in Figure 6.3(b), among other beautifully intricate polyhedra.

Exercise 6.5. *Find the ten Archimedean solids that are composed of just two distinct types of regular polygon faces, by generalizing equation (6.2).*

★ **Exercise 6.6.** *Prove that the dihedral angles of a regular polyhedron are identical.*

Although we have chosen not to delve deeply into higher dimensions in this book, we cannot resist a brief foray into the world of regular polytopes. A 3D polyhedron is composed of zero-dimensional vertices, one-dimensional edges, and two-dimensional faces. The generalization of a regular convex polyhedron to higher dimensions is a *regular convex polytope*. In four dimensions, these are composed of 3D regular polyhedral *facets* glued together in 4D space. Perhaps the most familiar is the *hypercube*, the 4D cube, composed of eight 3D cubes glued

(a)
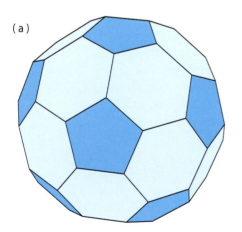

(b)
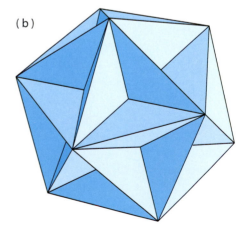

Figure 6.3. (a) Truncated icosahedron and (b) great dodecahedron.

together, square face to square face. Grasping the structure of any higher-dimensional object can be challenging. One method of display is to project a "wire-frame" version into 3D called the *Schlegel diagram* of the 4D polytope. Figure 6.4(a) provides such a projection of the hypercube, clearly showing seven of the eight cubes; the eighth surrounds the exterior of what is displayed. Figure 3.14 showed the Schlegel diagram of the 4D associahedron.

In contrast to the five Platonic solids, in 4D there are six regular polytopes. And in contrast to 4D, there are exactly three regular polytopes in each dimension $n \geq 5$: the n-simplex, generalizing the tetrahedron, the n-cube, generalizing the cube, and the n-orthoplex (or cross polytope), generalizing the octahedron. The three exceptional 4D regular polytopes are known as the 24-cell, the 120-cell, and the 600-cell. Figure 6.4(b) shows a Schlegel diagram of the 120-cell, composed of 120 dodecahedral facets. It was not until the nineteenth century that the list of regular polytopes was completed, approximately 2000 years after the three-dimensional regular polyhedra were constructed. Higher-dimensional polytopes have a surprising number of real-world applications and are the focus of active research today.

Exercise 6.7. *Draw Schlegel diagrams for the 4D tetrahedron (the 4-simplex) and the 4D octahedron (the 4-orthoplex).*

Exercise 6.8. *Find the number of 3D faces for the 4D associahedron whose Schlegel diagram was given in Figure 3.14.*

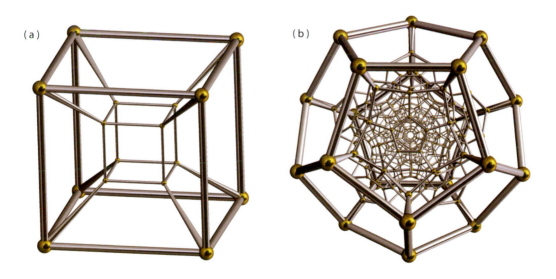

(a)   (b)

Figure 6.4. (a) The hypercube with Schläfli symbol {4, 3, 3} and (b) the 120-cell with Schläfli symbol {5, 3, 3}.

6.2 EULER'S POLYHEDRAL FORMULA

It took a surprisingly long time to define unambiguously what constitutes a single polyhedron. A big step in this clarification was made with Euler's formula, which we have encountered before as Theorem 3.12 in our study of triangulations, and which we revisit below. We now provide one definition of a polyhedron before turning to Euler's formula.

A *polyhedron* is composed of vertices, edges, and faces. We restrict each face to be a convex polygon to make the definition easier. To accommodate nonconvex polygonal faces, we simply partition these into (coplanar) convex pieces. These pieces are put together to form the surface of the polyhedron P subject to three conditions: the polygons intersect properly, the local topology is correct, and the global topology is correct. In more detail:

1. *Intersection condition.* The intersection of any two faces of P is either empty, a single vertex, or a single edge (and its endpoints).
2. *Local topology.* Every point p on the surface of P has a neighborhood (a small region surrounding and containing p) that is homeomorphic to an open disk.
3. *Global topology.* We would like the surface of P to be *connected*: a path exists on the surface between any two points of P. This excludes, for instance, a polyhedron with a floating internal cavity, but does not exclude a polyhedron with a tunnel that connects one part of the surface to another, as in Figure 6.5(c).

Let's expand a bit on the notion of a *homeomorphism*, which we encountered before and will encounter again. This is a term from topology intended to capture a notion of "topological equivalence." Two surfaces are *homeomorphic* if there is a continuous bijection (a homeomorphism) from one to the other whose inverse is also continuous. Loosely speaking, surfaces S_1 and S_2 are homeomorphic if S_1 can be cut

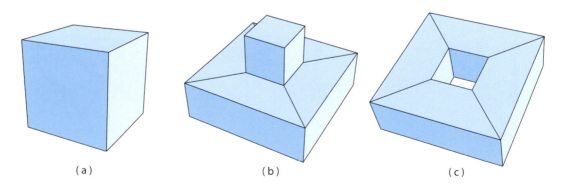

Figure 6.5. (a) A cube, (b) a polyhedral sphere, and (c) a polyhedral torus.

up into several pieces, each of which can then be continuously bent and stretched, and then glued back exactly along the seams of the same cuts, in order to form S_2. In terms of global topology, Figures 6.5(a) and (b) show examples of polyhedra that are homeomorphic to spheres — they are appropriately called *polyhedral spheres*.

Thus far, all examples of polyhedra we have seen in this book are homeomorphic to spheres. The object in Figure 6.5(c) is also a polyhedron, satisfying the three conditions above; however, it is homeomorphic *not* to a sphere but rather to a *torus* — the surface of a donut. One way to express the difference between a topological sphere and a torus is in the language employed in the Poincaré conjecture (Section 5.6): the sphere is *simply connected* in that every loop in the surface is contractible to a point, but a loop through a hole of the torus is not contractible. Just as this polyhedron has one hole, we could also have a two-holed, three-holed, or four-holed polyhedron, as pictured in Figure 6.6. Indeed, we can keep increasing the number of holes, creating an infinite number of polyhedra, none of which are homeomorphic to each other! The number of holes in a polyhedron is called the *genus* of the polyhedron.

In terms of local topology, we demand that the neighborhood of each point on the surface be homeomorphic to a disk. Two violations are shown in Figure 6.7. In part (a), a surface neighborhood of point p is homeomorphic to two disks touching at p — point p is surrounded by too much surface. The open box in part (b) is a *surface with boundary*, and a surface neighborhood of a point p on the boundary rim is homeomorphic to a half-disk — point p is not surrounded by enough surface.

Exercise 6.9. *Find a completely combinatorial definition of a triangulated polyhedron, that is, one that avoids using homeomorphisms.*

★ **Exercise 6.10.** *The definition of a homeomorphism is that it is a mapping between two spaces that (a) is one-to-one, (b) is continuous, and (c) has a continuous inverse. Construct examples to show that all three conditions are necessary to capture this notion of topological equivalence.*

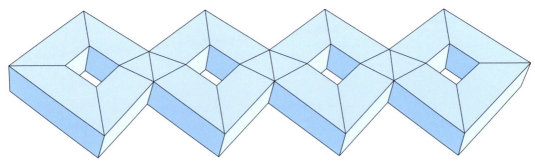

Figure 6.6. A polyhedron of genus four.

(a)   (b)

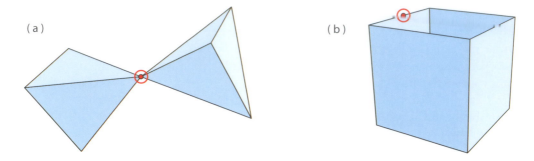

Figure 6.7. Two examples violating the local topology condition in the definition of a polyhedron.

One of the seminal events in the history of polyhedra was Leonhard Euler's discovery of his famous formula in 1750: the number of vertices and faces together is always two more than the number of edges, or $V + F = E + 2$. Although it is difficult to appreciate this advance from our perspective more than 250 years later, part of his achievement consisted in viewing a polyhedron as a combinatorial object rather than a purely geometric one. Once he isolated what a vertex, edge, and face should be, the regularity encapsulated by his formula became nearly obvious. However, he did not settle on a precise definition of a polyhedron, and partly as a consequence, his attempted proof was incomplete. The first rigorous proof was provided by Adrien-Marie Legendre in 1794. And even then the exact scope of the formula's applicability, which turns on the definition of a polyhedron, followed a long and tangled history. Indeed, the great dodecahedron in Figure 6.3(b) was initially not considered a regular polyhedron because treating it as composed of 12 intersecting pentagonal faces (as Johannes Kepler did) violates Euler's formula.

Exercise 6.11. *Compute $V - E + F$ for the great dodecahedron in two ways: (a) when viewed as composed of many congruent triangular faces, and (b) when viewed as composed of interpenetrating congruent regular pentagons.*

Today, dozens of distinct proofs of Euler's formula are recognized. We already detailed one in Chapter 3 in the context of planar graphs. Here we present one of the prettiest proofs, due to Karl von Staudt from 1847, which has the dual advantages of being explicitly based on polyhedra and employing spanning trees, an important concept we will have occasion to use later.

Theorem 6.12 (Euler). *For any polyhedron homeomorphic to a sphere with V vertices, E edges, and F faces, V − E + F = 2.*

Proof. The proof is partitioned into five steps.

1. Convert the 1-skeleton of the polyhedron to a plane graph G.
2. Select a spanning tree T of G.
3. Construct the dual graph G^* of G.
4. Identify the complementary spanning tree T^* of G^*.
5. Apply a tree counting lemma to both T and T^*.

We now expand and illustrate each step below.

1. *Polyhedron 1-skeleton to plane graph* G: The *1-skeleton* of a polyhedron P is the graph G of vertices and edges on its surface. We want to embed G in the plane with noncrossing arcs. This can be viewed as flattening G to the plane, which can be accomplished as follows: Choose an arbitrary face f of P and remove it, leaving a hole in the surface. Now stretch the hole wider and wider until it becomes much larger than the original size of P. We can then flatten P, and therefore G, into the plane in such a way that the (possibly distorted) arcs of G deriving from edges of P do not cross.[1]

   This process is illustrated for the dodecahedron in Figure 6.8(a). Note that the outer boundary of G bounds the *exterior face* of G, corresponding to the removed f. So G has V vertices, E edges, and F faces — the same counts as P. The remainder of the proof uses this plane graph G.

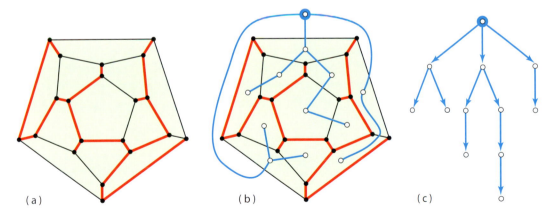

(a)   (b)   (c)

Figure 6.8. (a) Spanning tree T of the Schlegel diagram G for the dodecahedron. (b) Complementary spanning tree T^* of dual G^*. (c) A one-to-one correspondence between edges and nonroot vertices of a tree; the tree is T^* in this case.

---

[1] This planar representation of the 1-skeleton is the Schlegel diagram for the 3D polyhedron.

2. *Spanning tree T of G*: A *tree* is a connected graph without cycles. A *spanning tree* of a graph G is a tree formed by a subset of the edges of G such that every vertex of G is incident to some edge of T. We say that T *spans* the vertices of G. There are in general many spanning trees for a given graph and they are not difficult to find. Let T be any one, as drawn in red in Figure 6.8(a).

3. *Dual graph G**: The *dual* G^* to a plane graph G has a node for each face of G, and connects two nodes by an arc if the corresponding faces of G share an edge. (We encountered this notion in Section 4.3, where we showed that the dual of the Voronoi diagram is the Delaunay triangulation.) It is important in general (and in our context) to include the exterior face of G. The faces of G become vertices of G^*, and the edges of G and G^* are in one-to-one correspondence. It is also not difficult to see that the faces of G^* correspond to vertices of G.

4. *Complementary spanning tree T* of G**: Let T^* be the subgraph of G^* whose arcs correspond to edges of G that are not part of T, that is, the edges of $G \setminus T$. Figure 6.8(b) shows the tree T^* in blue. Because T has no cycles, T^* connects all the faces of G; in other words, all the faces of G are accessible via paths in G^* from the exterior face without crossing an edge of T. So T^* spans the vertices of G^*. But now notice that T^* cannot itself contain a cycle, for this would separate some vertex of G inside the cycle from those outside, which would contradict the fact that T is spanning and that T and T^* do not cross. So T^* is a spanning tree[2] of G^*.

5. *Apply lemma to both T and T**: The lemma is simply that, if a tree has n vertices, it must have exactly $n-1$ edges. This can easily be established by induction (Exercise 6.14), but the following proof is perhaps even more straightforward. Pick any vertex as the root of the tree and direct each edge away from the root, pointing to its children vertices. Now each directed edge points to a nonroot vertex, establishing a one-to-one correspondence between edges and nonroot vertices; Figure 6.8(c) shows an example. With only the root node without a partner, there must be $n-1$ edges.

   Since G has V vertices, and because T spans G, T also has V vertices. By the lemma, T has $V-1$ edges. Since G^* has F vertices (one per face of G), and because T^* spans G^*, T^* also has F vertices. By the lemma, T^* has $F-1$ edges.

   Thus the number of edges of T and T^* totals $(V-1)+(F-1)$. But by the construction of T^*, it contains edges corresponding exactly to all those edges that are not in T, so together they account for all E edges. Thus we have established that $V-E+F=2$.   □

---

[2] Sometimes T and T^* are called *interdigitating* trees.

The remarkable consequence of this theorem is that regardless how a polyhedral sphere is made, Euler's formula will always result in $V - E + F$ equaling 2. Let's consider some examples of polyhedra (which are homeomorphic to spheres) we have seen. Figure 2.13 displays a polyhedron with 758 vertices, 2268 edges, and 1512 triangular faces. Substituting these values into the formula yields 2 as the answer. The 3D associahedron pictured in Figure 3.13(b) has $V = 14$, $E = 21$, and $F = 9$. Once again we see $V - E + F = 2$. Indeed, our examples do not need to be convex polyhedra because the formula holds for *any* polyhedron homeomorphic to a sphere. The example of Figure 6.5(b) shows a nonconvex polyhedron with $V = 16$, $E = 26$, and $F = 12$, satisfying $V - E + F = 2$.

Exercise 6.13. *Form a polyhedron P by gluing k dodecahedra together in a string, each one sharing a pentagon face with its neighbors. Remove all those internal shared pentagons and verify Euler's formula for P.*

Exercise 6.14. *Using induction, prove the tree lemma: a tree of n vertices has $n - 1$ edges.*

We restricted attention in the proof of Euler's formula to polyhedra homeomorphic to a sphere, having no holes. This assumption was essential to the first step of the proof: one cannot flatten the 1-skeleton of a polyhedron with a hole by the removal of a single face. However, a generalization of Euler's formula does hold for polyhedra with holes. As we stated earlier, the number of holes in a polyhedron is known as its *genus*. Finding the genus of a polyhedron is by no mean obvious (see Figure 6.10), but we will see that this is handled by the Euler formula generalization.

Although Euler created his formula for polyhedra, it actually measures a topological invariant of surfaces more generally constructed. It would be a long diversion into topology to formally define every concept we need to describe this carefully, so we content ourselves with an intuitive sketch of the main ideas. We take as a *surface S* anything homeomorphic to a polyhedron.[3] In order to apply Euler's formula to surface S, we draw a connected graph G on it, partitioning S into vertices, edges, and faces — we say G is *embedded* on S. So the edges of G can be arbitrary curves that meet each other only at vertices, and the faces are all homeomorphic to polygons. When all the faces of the graph are triangles, the surface is said to be *triangulated* by G — more generally the surface is *meshed* by the graph. (For example, quadrilateral meshes, where each

---

[3] All the surfaces we consider are assumed to be *orientable* (roughly, those with two sides).

face is a quadrilateral, are commonly employed in architectural design.) Of course, one could draw an infinite number of different graphs on the same surface.

Definition. For a surface S partitioned by an embedded graph G into V vertices, E edges, and F faces, the *Euler characteristic* $\chi(S)$ of S is $V - E + F$.

From this definition, it seems the Euler characteristic of the surface is clearly dependent on the embedded graph G which partitions it. The following theorem makes the remarkable claim that Euler's formula yields the exact same result independent of the partition! Moreover, it says the Euler characteristic depends only on the genus of the surface. So, finally, here is the generalization of Euler's formula to arbitrary surfaces, first achieved by Simon L'Huilier in 1813 (his name translates as "the oiler").

Theorem 6.15 (Euler). *For a surface S of genus g, $\chi(S) = 2 - 2g$.*

Sketch of Proof. We aim to show that for any embedded graph G partitioning a surface S of genus g, the equation $V - E + F = 2 - 2g$ holds. The proof is by induction on g. For $g = 0$, the surface S is homeomorphic to a sphere, and Theorem 6.12 applies to show that $\chi(S) = 2$.

Assume the theorem holds for all values of genus less than g, and consider a surface S with genus g. Let S be meshed by a graph G. Choose a loop λ on G that does not disconnect the surface. If no such loop exists, the surface must be a sphere and we are done. Now separate S along λ, capping the two loop copies with two new faces, as illustrated in Figure 6.9. This creates a new surface S' of genus $g - 1$. Note that the separation doubles the number of edges and vertices along λ, again not altering the value of $\chi(S)$. However, the addition of the two new faces implies $\chi(S') = \chi(S) + 2$. But by the induction hypothesis, we see

$$\chi(S) + 2 = \chi(S') = 2 - 2(g - 1) = 2 - 2g + 2,$$

and the result follows. □

Notice that the statement and the proof of the theorem are independent of the partition by the graph G. Indeed, the work needed to show this was hidden within the proof of Theorem 6.12, which handled the base case $g = 0$. Thus the Euler characteristic $\chi(S)$ yields exactly the same result for any connected G, depending only on the genus of S. Moreover, remarkably, we get a *global* property — the genus — from a purely *local* accounting of V, E, and F. The Euler characteristic has turned out to be a key invariant in the study of general topological spaces: it is the alternating sum of the *Betti numbers*, which record the rank of the

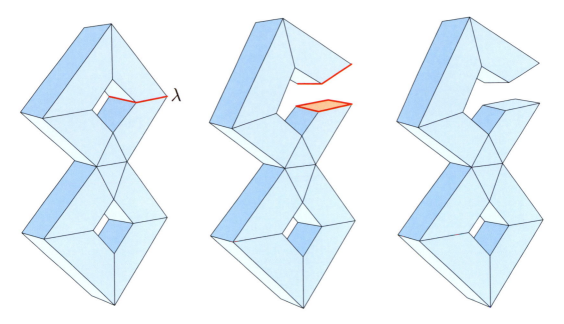

Figure 6.9. Slice along a curve λ and attach two new faces to decrease the genus of the surface.

homology groups for the space. We will not have occasion to explore further in this direction, but the Euler characteristic will play a role in the proof of the Gauss-Bonnet theorem in Section 6.3 below.

Exercise 6.16. *Verify Euler's formula for the polyhedron in Figure 6.6.*

Exercise 6.17. *Compute the Euler characteristic and genus of the two polyhedra shown in Figure 6.10.*

Exercise 6.18. *Although we insisted that the surface of a polyhedron be connected, Euler's formula applies to more general surfaces. Compute the Euler characteristic of a cube with a cubical cavity. Would you conjecture that this characteristic holds for any topological sphere with a spherical cavity?*

From these examples, one might wonder whether Theorem 6.15 implies that the Euler characteristic must always be an even integer. The answer is YES for all closed (orientable) surfaces, but it can be odd for surfaces with boundary.

Exercise 6.19. *What is the Euler characteristic of a topological disk?*

Exercise 6.20. *Compute the Euler characteristic for the open box in Figure 6.7(b).*

(a)

(b)

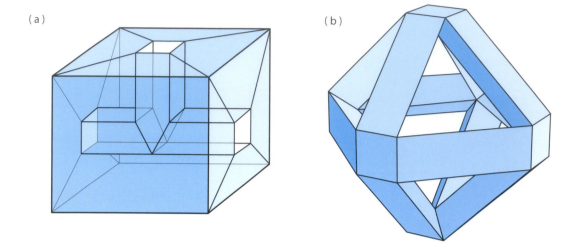

Figure 6.10. Find the genus of each polyhedron.

Exercise 6.21. *Compute the Euler characteristic of a cylindrical band.*

Exercise 6.22. *Conjecture and prove a generalization of Theorem 6.15 which extends to surfaces with boundary.*

6.3 THE GAUSS-BONNET THEOREM

We have emphasized that Euler's combinatorial formula is a topological invariant of surfaces, including polyhedral surfaces. In some sense, there is no geometry in Euler's formula. Perhaps the most far-reaching theorem concerning the geometry of surfaces is the *Gauss-Bonnet theorem*, one of the jewels of differential geometry. In that field, it is often phrased in the following terms:

Theorem 6.23 (Gauss-Bonnet). *Let S be a smooth surface without boundary. Then*

$$\int_S K \, dA = 2\pi \, \chi(S). \tag{6.3}$$

It will take some time to explain this formula and specialize it to polyhedral surfaces, but the time is well spent. We start by identifying the symbols in equation (6.3). Here K is the *Gaussian curvature*, an intrinsic (within the surface) measure of how sharply S is curved at each point, the analog of κ from Section 5.5 but for surfaces rather than curves. This is a *signed* measure, with positive curvatures indicating protrusions and dents, negative curvatures indicating saddle points, and zero curvatures implying flatness. Figure 6.11 shows a point on different surfaces having

Figure 6.11. Surfaces of positive, zero, and negative curvatures.

positive, zero, and negative curvatures, respectively. Given that dA is the area differential, the first integral sums the curvature over the whole surface S.

So why is this a remarkable theorem? The left side of the equation deals with curvature, a purely geometric concept. The right side deals with the Euler characteristic, a purely topological concept. Gauss-Bonnet says that these two ideas are fundamentally linked. From the left side of the equation, it claims that if you know the (local) curvature at every point on the surface, the global shape of the surface (its genus) is determined. From the right side of the equation, it says that if the genus of the surface is given, the curvatures must add up to a constant. In other words, if we dent the surface, saddles must emerge so that the amount of negative and positive curvatures cancel out perfectly! For instance, in Figure 6.12, even with all the twists and turns creating patches of negative curvature, the total curvature of the surface is exactly 4π, as with any surface homeomorphic to a sphere.

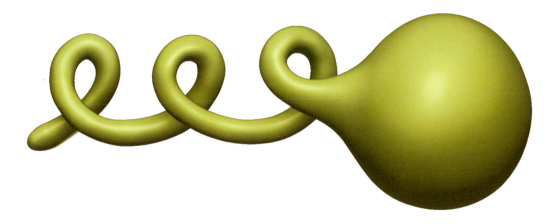

Figure 6.12. A deformation of the sphere whose total Gaussian curvature is 4π. Figure courtesy of Matthew Harvey.

Carl Friedrich Gauss, arguably the greatest mathematician since antiquity, knew the Gauss-Bonnet theorem (in one of its differential geometry versions) by 1825, but did not publish a proof, which was supplied by Pierre Bonnet in 1848 and subsequently generalized in several directions by many others. The descendants of this seminal theorem include the Riemann-Roch theorem and the Atiyah-Singer index theorem, the latter of which plays a central role today in theoretical physics.

To consider the polyhedral version of Gauss-Bonnet requires an understanding of *discrete curvature*.

Definition. The *Gaussian curvature* $K(p)$ at a point p of a polyhedral surface is 2π minus the sum of the face angles incident to p.

This is sometimes called the *angle deficit* at a vertex. If the surface is cut open and flattened in the neighborhood of v, there will be an angle gap of precisely $K(v)$. Thus a point p in the interior of a face of P has curvature zero since the sum of angles around p sum to 2π. Less obvious is that a point on a polyhedral edge e also has curvature zero, for it has π face angle incident from either side of e. Thus *all* the curvature of a polyhedron is concentrated at its vertices. Figure 6.13 shows examples of polyhedral curvatures.

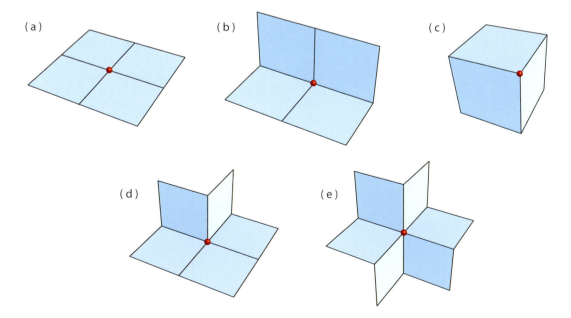

Figure 6.13. Polyhedral curvatures 0, 0, $\pi/2$, $-\pi/2$, and $-\pi$ at the marked points, respectively.

Exercise 6.24. *What is the curvature at each vertex of a dodecahedron? What is the sum of those curvatures over all vertices?*

With the notion of discrete curvature defined, we may now phrase the polyhedral version of the Gauss-Bonnet theorem:

Theorem 6.25 (Polyhedral Gauss-Bonnet). *For a polyhedron P,*

$$\sum_{v \in P} K(v) = 2\pi \, \chi(P). \qquad (6.4)$$

Proof. By the definition of $K(v)$, we have

$$\sum_{v \in P} K(v) = \sum_{v \in P} \left(2\pi - \sum_{(f \text{ incident } v)} f \text{ angle at } v \right)$$

$$= \sum_{v \in P} 2\pi - \sum_{v \in P} \left(\sum_{(f \text{ incident } v)} f \text{ angle at } v \right). \qquad (6.5)$$

The first sum of equation (6.5) is just $\sum_{v \in P} 2\pi = 2\pi V$. The second sum adds up all the face angles incident to v, over all vertices v in P. So this is just the sum of all face angles over the surface, each counted exactly once. Let n_f be the number of sides of the polygonal face f. Then the sum of the face angles of f is $(n_f - 2)\pi$. So we have

$$\sum_{v \in P} \left(\sum_{(f \text{ incident } v)} f \text{ angle at } v \right) = \sum_{f \in P} (n_f - 2)\pi$$

$$= \pi \left(\sum_{f \in P} n_f - \sum_{f \in P} 2 \right). \qquad (6.6)$$

In equation (6.6), $\sum_{f \in P} 2 = 2F$ and $\sum_{f \in P} n_f = 2E$, the latter because each edge is shared by two faces, so summing n_f over all f double-counts the edges of P. Returning to equation (6.5), we have

$$\sum_{v \in P} K(v) = 2\pi V - \pi(2E - 2F) = 2\pi \chi(P),$$

which establishes the Gauss-Bonnet theorem in the form of equation (6.4). $\qquad \square$

The first polyhedral version of the Gauss-Bonnet theorem was discovered by René Descartes in 1630, predating Euler's formula by two decades, and predating the full Gauss-Bonnet theorem by two centuries. Descartes only considered polyhedral spheres, and phrased the result in his *Treatise on Polyhedra* saying that

"…in a solid body all the exterior angles, taken together, equal eight solid right angles."

Here the "eight solid right angles" constitute 4π, the value of the right side of equation (6.4) when $\chi(S) = 2$ is a polyhedral sphere.

Exercise 6.26. *Verify the Gauss-Bonnet formula for the three polyhedra shown in Figure 6.5.*

Exercise 6.27. *Verify the Gauss-Bonnet formula for the polyhedron shown in Figure 6.10(a).*

The Gauss-Bonnet theorem has an important generalization to surfaces S with boundary ∂S, such as that in Figure 6.7(b). We will use this generalization in Section 6.6. First we state the theorem for smooth surfaces and then present a discrete version.

Theorem 6.28 (Gauss-Bonnet). *Let S be a surface and ∂S its boundary. Then*

$$\int_S K\, dA + \int_{\partial S} k_g\, ds = 2\pi\, \chi(S)\,. \tag{6.7}$$

The symbol k_g is the *geodesic curvature* at a point on the curve C. This curvature k_g measures the deviation of a curve from straightness within the surface tangent plane at a point of C. The second integral sums the geodesic curvature over the boundary curve ∂S; here ds is the arc-length differential. So Theorem 6.23 is the special case of this theorem for surfaces without boundary, where the second integral term in equation (6.7) is zero.

In the discrete polyhedral setting, the notion of Gaussian curvature needs to be extended to polyhedra with boundary:

Definition. Let P be a polyhedron with boundary. The *geodesic curvature* $K(p)$ at a boundary point p of P is π minus the sum of the face angles incident to p.

This discrete analog of the geodesic curvature $k_g(p)$ is the *turn angle* at each p on a boundary curve of P: the signed angle turn experienced by a person walking along the boundary while vertically aligned with the surface normal. This turn is zero at every point interior to a straight segment of the boundary since the sum of the face angles is exactly π. Thus the geodesic curvature is solely concentrated at the corners of the boundary curves. The discrete version of Theorem 6.28 now follows:

Theorem 6.29 (Polyhedral Gauss-Bonnet). *Let P be a polyhedron, ∂P be its boundary, and P \ ∂P be its interior. If* $\omega = \sum\limits_{v \in P \setminus \partial P} K(v)$ *and*

$\tau = \sum\limits_{v \in \partial P} K(v)$, *then*

$$\omega + \tau = 2\pi \, \chi(P). \tag{6.8}$$

In words: the sum of the curvature in the interior of the surface P plus the sum of the curvature along its boundary is a constant dependent only on the Euler characteristic of P.

Exercise 6.30. *Provide a proof of Theorem 6.29, tracking the proof of Theorem 6.25.*

There is another way of understanding this equation for polyhedra *without* boundary. Let C be a counterclockwise-oriented simple closed curve on the polyhedron P. The sum of the curvature of the region *bounded* by C—the portion of P to the left of C—is the first summation term ω, whereas the sum of the geodesic curvature *along* the curve C is the second summation term τ. Here the curve C on the polyhedron P without boundary plays the role of ∂P.

Example 6.31. Figure 6.14 shows three examples of curves C on a cube. The curve in part (a) is obtained by slicing the cube horizontally with a plane. There are four corners on this curve, where the curve crosses an edge of the cube. The geodesic curvature at each of these corners is zero since the sum of the face angles to the left side of C is π. Alternatively, the person walking along the curve does not turn with respect to the normal.

The curve in part (b) has four corners as well, obtained by slicing the cube with a diagonally slanting plane. Here the geodesic curvature

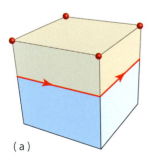
(a)

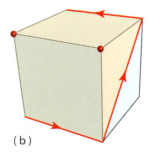
(b)

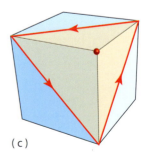
(c)

Figure 6.14. The three curves turn leftward 0, π, and $3\pi/2$, respectively, and so enclose 2π, π, and $\pi/2$ by equation (6.8).

at each of the corners is now $\pi/4$: there is $\pi/2 + \pi/4$ to the left of the walker at each corner, so they must deviate leftward from straightness by $\pi - (\pi/2 + \pi/4) = \pi/4$. In part (c), there are three corners of C, each with turn angle equal to $\pi/2$.

Equation (6.8) expresses a tradeoff between geodesic curvature (turn angle) and enclosed curvature. If C surrounds a vertex-free region of a polyhedral sphere P, then $\omega = 0$ and so $\tau = 2\pi$, which is the familiar result that total turn around a planar curve is 2π. We saw this both in Exercise 1.12 and in our discussion of the winding number in Section 5.4, and we return to this topic in Section 6.6.

Example 6.32. Continuing the discussion from Example 6.31, we see that the curve C in Figure 6.14(a) does not turn at all. So $\tau = 0$ and there must be $\omega = 2\pi$ curvature to each side of C, which makes sense: The total curvature of 4π (for the cube) is equally distributed to each side of the curve having four vertices (each vertex carrying $\pi/2$ Gaussian curvature). Notice the more C turns leftward, the less curvature it encloses: part (b) has $\tau = 4(\pi/4)$, implying $\omega = \pi$, capturing two vertices. Finally, Figure 6.14(c) shows a curve where $\tau = 3(\pi/2)$ and $\omega = \pi/2$, surrounding one vertex.

Exercise 6.33. *Verify equation (6.8) when the polyhedron is a regular tetrahedron T and C is the curve of intersection of a plane parallel to the base of T, midway between the base and the apex of T.*

Exercise 6.34. *Verify equation (6.8) for the two curves shown in Figure 6.15.*

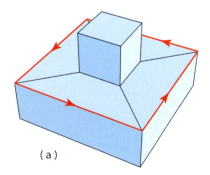
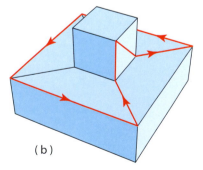

(a)   (b)

Figure 6.15. Closed oriented curves on the surface of a nonconvex polyhedron.

6.4 CAUCHY RIGIDITY

Both Euler's formula and the Gauss-Bonnet theorem hold for all varieties of polyhedra, whether they be convex or nonconvex, with or without holes, with or without boundary. For the remainder of the chapter we concentrate on convex polyhedra, for which several beautiful theorems hold which may fail for nonconvex polyhedra. In this section we describe and prove a rigidity theorem for polyhedra, which relies at a crucial juncture on Euler's formula. In fact, Augustin Cauchy, the profound French mathematician, presented the 1813 proof of his rigidity theorem in the same paper in which he offered a proof of Euler's formula.

The problem solved by Cauchy goes back to Euclid, who defined equality between polyhedra (*Elements*, Book XI) as follows:

"Equal and similar solid figures are those contained by similar planes equal in multitude and magnitude."

In modern terminology, it states that two polyhedra are congruent if they have congruent faces similarly arranged about each vertex. This is certainly false for nonconvex polyhedra, as the example in Figure 6.16 demonstrates. Here the roof over the cube can dent inward as well as protrude outward while retaining the congruency of the faces and their arrangements about each vertex. But Euclid was probably thinking only of convex polyhedra. Still, it was realized in the nineteenth century that Euclid's definition of equivalent polyhedra was actually a theorem in need of a proof, a proof Cauchy supplied. One phrasing of his theorem is as follows:

Theorem 6.35 (Cauchy Rigidity). *If two closed, convex polyhedra are combinatorially equivalent, with corresponding faces congruent, then*

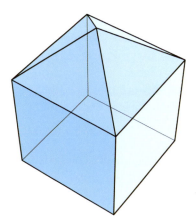
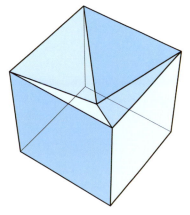

Figure 6.16. Two different polyhedra with congruent faces similarly arranged about each vertex.

the polyhedra are congruent; in particular, the dihedral angles at corresponding edges are the same.

The reason it is called Cauchy's *rigidity* theorem is that it implies that a convex polyhedron is *rigid* in the sense that, if it were built with face plates hinged along edges, it could not flex. Cauchy's theorem is, however, strictly stronger than claiming that convex polyhedra are rigid, for it is conceivable that they are rigid but there are several different, isolated, incongruent realizations. Cauchy's theorem says that even this is not the case: There must be a unique realization. The proof of this theorem is by contradiction consisting of three main steps, each highly original and much used subsequently:

1. A geometric lemma now known as Cauchy's arm lemma.
2. A combinatorial lemma on sign changes on planar graphs.
3. A proof by contradiction employing these two lemmas.

Cauchy's arm lemma is a surprisingly delicate result that has proved useful in many contexts aside from the role it plays in his rigidity theorem. In one sense it is obvious: it claims that straightening a convex chain increases the distance between its endpoints. But its subtlety is indicated by the fact that Cauchy's own proof of this lemma was flawed, and the flaw was not noticed for over a century, when Ernst Steinitz repaired it. For this reason it is sometimes known as the Cauchy-Steinitz lemma.

Although the arm lemma needed for Cauchy's rigidity theorem concerns convex chains composed of geodesic segments on a sphere, the proof in the plane differs from that on the sphere in only minor ways, and we choose here to avoid the (minimal) spherical trigonometry needed and only present the planar version of the lemma.

Let $C = (v_0, v_1, \ldots, v_n)$ be a planar convex polygon. We will view C as an open chain, missing the last edge $v_n v_0$, and call it a *planar convex chain*. The other edges of C are considered rigid bars connected by joints at the interior vertices $v_1, \ldots, v_{n-1}$. The internal angles $\alpha_1, \ldots, \alpha_{n-1}$ at the joints all lie in the range $(0, \pi]$. Although the joints are free to rotate, the bars always retain their original length. With this notation, here is one phrasing of Cauchy's lemma:

Lemma 6.36 (Cauchy's Arm). *If a planar convex chain C is opened by increasing some or all of its internal angles, but not beyond π, then the distance between v_0 and v_n is strictly increased.*

In other words, if the internal angles α_i are replaced by α_i', with $\alpha_i \leq \alpha_i' \leq \pi$, then $|v_0' - v_n'| > |v_0 - v_n|$. It is important to realize that an open polygonal chain, all of whose internal angles $\alpha_1, \ldots, \alpha_{n-1}$ are convex, is

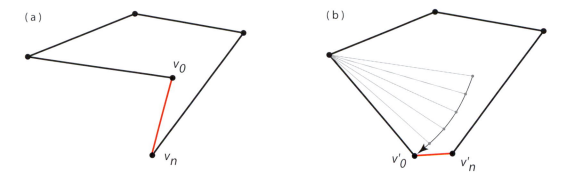

Figure 6.17. (a) A chain that fails to be convex (b) results in a contradiction to Cauchy's arm lemma.

not necessarily a convex chain as defined above: it must be that connecting v_0 to v_n produces a convex polygon (in particular, α_0 and α_n must be convex as well). Indeed, the lemma could be false otherwise, as illustrated in Figure 6.17. Part (a) shows a chain whose internal angles are convex. As an internal angle increases, the distance between the endpoints actually decreases, pictured in (b). In fact, the situation shown in this figure could arise at a substep in Cauchy's original induction proof, which was the flaw repaired by Steinitz. Here we will present the beautiful non-inductive proof found by Stanislaw Zaremba in 1967.

Proof. Let $C = (v_0, v_1, \ldots, v_n)$ be a planar convex chain. Establish a coordinate system in the plane of the chain as follows. The x-axis contains v_0 and v_n, with v_n right of v_0. The y-axis passes through the vertex v_k furthest from the x-axis. (If there are two vertices tied for furthest, let v_k be the right one.) Chain C is opened to a new chain $C' = (v'_0, v'_1, \ldots, v'_n)$, and placed so that:

1. Vertex v_k does not move: $v'_k = v_k$.
2. The new angle $v'_{k-1} v_k v'_{k+1}$ contains the old angle $v_{k-1} v_k v_{k+1}$.
3. Neither v'_{k-1} nor v'_{k+1} is placed above v_k.

Because $\alpha_k \leq \alpha'_k \leq \pi$, it is possible to satisfy these three conditions, as demonstrated in Figure 6.18. If we define x_i to be the x-coordinate of v_i, then $|v_n - v_0| = x_n - x_0$ because of the choice of coordinate system. The plan is to compute $x'_n - x'_0$, which is a lower bound on $|v'_n - v'_0|$.

For the right portion of the chain, from v_k to v_n, define θ_i to be the angle between the ray $v_i v_{i+1}$ and the positive x-axis. This angle lies in $[-\pi, 0)$, for all these rays point downward. The opening motion increases (or leaves unchanged) each θ_i for each ray $v_i v_{i+1}$ is turned counterclockwise about v_i by opening angles. The new angles never exceed 0 because v'_{k+1} is below v_k, so we have $\theta_i \leq \theta'_i \leq 0$.

Let ℓ_i be the length of the edge $v_i v_{i+1}$. Then the x-coordinate x_n of v_n can be computed as the sum of the horizontal contributions of

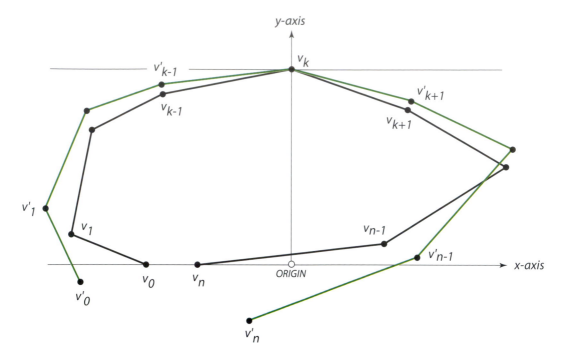

Figure 6.18. Opening C while keeping v_k fixed and highest.

each edge:

$$x_n = \ell_k \cos \theta_k + \ell_{k+1} \cos \theta_{k+1} + \cdots + \ell_{n-1} \cos \theta_{n-1}. \qquad (6.9)$$

Each $\theta_i' \geq \theta_i$ and because the cosine is an increasing function of its argument in the range $[-\pi, 0]$, it must be that $x_n' \geq x_n$; in other words, v_n moves rightward. By symmetry, a similar argument establishes that v_0 moves leftward: $x_0' \leq x_0$. By assumption, at least some angles open, which means that at least one of the two inequalities is strict. Thus we have

$$|v_n' - v_0'| \geq x_n' - x_0' > x_n - x_0 = |v_n - v_0|,$$

which proves our claim.   □

Exercise 6.37. *Suppose $v_0 = v_n$ initially. Is Cauchy's arm lemma still valid?*

Exercise 6.38. *The condition $\alpha_i \leq \alpha_i' \leq \pi$ in Lemma 6.36 may be rewritten $\pi - \alpha_i \geq \pi - \alpha_i' \geq 0$. The angle $\pi - \alpha_i$ is the* turn angle *at the joint. Suppose the condition is generalized to $\pi - \alpha_i \geq |\pi - \alpha_i'| \geq 0$, that is, the new turn angle in absolute value is no larger than the old. Is the conclusion $|v_0' - v_n'| > |v_0 - v_n|$ still valid? Explore enough to make a conjecture.*

Exercise 6.39. *Continuing the previous exercise, consider reconfiguring the planar convex chain C into a new chain C' in $\mathbb{R}^3$. Define the turn angle τ_i' at v_i' to be arccos of the dot product of unit vectors along $v_{i+1} - v_i$ and $v_i - v_{i-1}$, so that $\pi > \tau_i' \geq 0$. Suppose the angle condition in Cauchy's lemma is generalized to $\pi - \alpha_i \geq \tau_i'$. Is the conclusion $|v_0' - v_n'| > |v_0 - v_n|$ still valid? Explore enough to make a conjecture.*

The second piece of the puzzle we need to prove the rigidity theorem is a condition on sign alternations of planar graphs. It is this lemma that employs Euler's formula at its core, applied to any connected plane graph as proved in Theorem 3.12. The lemma concerns a plane graph G whose edges have been 2-colored, that is, colored (arbitrarily) with two distinct colors, say red and blue. As we (cyclically) walk around the edges of a vertex v, the lemma counts the color changes of the edges incident to v. For example, {red, red, blue, red} represents two color changes.

Lemma 6.40. *Let G be a plane graph with edges that are 2-colored. Then there is a vertex v of G with at most two color changes in cyclic order around v.*

Proof. Suppose to the contrary that at every vertex v of G, there are more than two color changes around v. The number of color changes about a vertex must be even because each blue to red transition must eventually be followed by a red to blue transition. So if there are more than two, there must be four or more.

We can associate each color change about v with a unique face angle incident to v, the face bounded by the two edges with different colors. Let's call a face angle representing a color change a *cc-corner*. Figure 6.19 shows a 2-colored plane graph illustrating cc-corners. Let G have V vertices, E edges, and F faces, and let c be the total number of cc-corners in G. Then, because each vertex has at least four color changes by hypothesis, we know $c \geq 4V$.

Having counted c based on vertices, let's now count it in terms of faces. Let a face f have k edges. If k is even, it could be that all k face angles are cc-corners. If k is odd, at most $k - 1$ cc-corners can occur. Let f_k be the number of faces in G bound by k edges. Then, toward the value of c, f_3 contributes at most $2 f_3$, f_4 contributes at most $4 f_4$, and so on. So we get an upper bound on c, where

$$4V \leq c \leq 2 f_3 + 4 f_4 + 4 f_5 + 6 f_6 + 6 f_7 + \cdots$$
$$\leq 2 f_3 + 4 f_4 + 6 f_5 + 8 f_6 + 10 f_7 + \cdots$$
$$= 2(3 f_3 + 4 f_4 + 5 f_5 + 6 f_7 + \cdots) - 4(f_3 + f_4 + f_5 + f_7 + \cdots)$$
$$= 2(2E) - 4(F) = 4E - 4F,$$

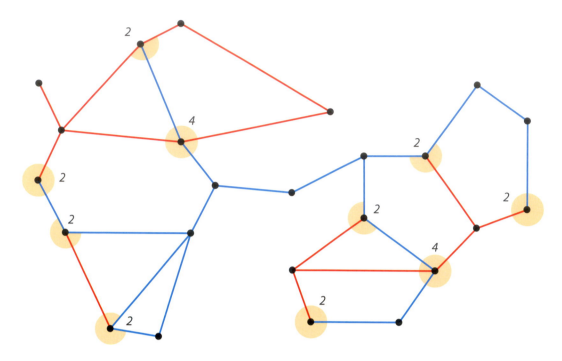

Figure 6.19. A 2-colored plane graph with the face angles representing cc-corners marked in orange, and the number of cc-corners incident to those vertices indicated.

since $\sum k f_k$ double counts each edge and $\sum f_k$ counts each face once. Thus $4V \leq 4E - 4F$, implying $V - E + F \leq 0$ for a planar graph G. This contradicts Euler's formula (Theorem 3.12) and so establishes the lemma. □

We now complete the proof of Cauchy's rigidity theorem using contradiction.

Proof of Cauchy's Rigidity Theorem. Let P and P' be two incongruent convex polyhedra that are combinatorially equivalent, with corresponding faces congruent. Because their combinatorial structure is the same, their vertices, edges, and faces can be matched one-to-one.

Color each edge e of P with blue or red if the dihedral angle at e is larger or smaller, respectively, than its corresponding edge of P'. If the angle is the same, then apply no color to the edge. Let G be the subgraph of the 1-skeleton of P containing the colored edges. Because the two polyhedra are incongruent, some edges must be colored, and so G is nonempty. However, it may consist of several disconnected components. Let H be one such component. Apply Lemma 6.40 to the plane graph H to conclude that there is at least one vertex v with at most two color changes in the cyclic order of the edges of H incident to v. Our goal is now to show that the geometry in a neighborhood of v in P violates Cauchy's arm lemma.

Let v' in P' be the vertex corresponding to v in P. Intersect P with a small sphere S_v centered on v. Each face f incident to v intersects S_v in an arc of a great circle: the plane containing f passes through the center of S_v and so intersects S_v in a great circle, and restricting to f selects an arc of this circle whose length is proportional to the face angle of f at v. Therefore, the collection of faces of P incident to v intersects S_v in a convex spherical polygon $Q = P \cap S_v$. Figure 6.20 shows two views of a spherical polygon produced by intersecting a polyhedron in the neighborhood of a vertex v with a sphere S_v centered on v. What is crucial to realize is that the angle at a vertex p of polygon Q is equal to the dihedral angle of the edge e in P that penetrates S_v at p. In other words, the dihedral angle of a 3D polyhedron is converted to a vertex angle of a 2D polygon!

Now we compare Q with its corresponding spherical polygon $Q' = P' \cap S_{v'}$. Because the faces of P and P' are congruent, the arcs of Q and Q' have the same length. But the selection of v from H ensures that some dihedral angles are colored and so unequal. Now we consider the two possibilities resulting from Lemma 6.40: There are either no or two color changes about v.

1. *There are no sign changes.* Without loss of generality, say that all edges of H incident to v are labeled blue. The other edges of P incident to v must then be uncolored. So the spherical polygon Q must have blue vertices, and possibly some vertices with no color, as pictured in Figure 6.21(a). (Although we are dealing with spherical polygons as

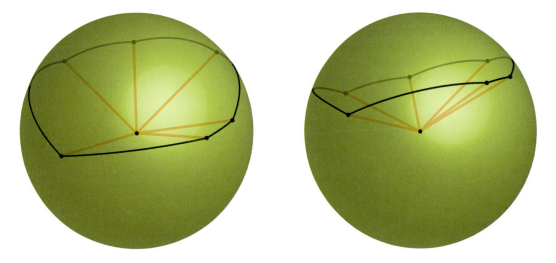

Figure 6.20. Two views of a spherical polygon produced by intersecting a polyhedron in the neighborhood of a vertex v with a sphere S_v centered on v.

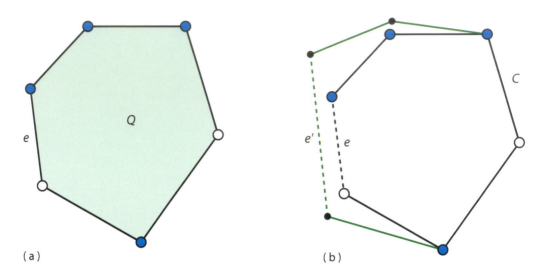

(a)   (b)

Figure 6.21. (a) The spherical polygon Q having no sign changes, with vertices colored blue or with no color. (b) The chain C opens up due to the blue vertices, showing an increase in the length of an edge, resulting in a contradiction.

depicted in Figure 6.20, we represent them as Euclidean polygons as the arguments apply naturally to both types.) Recall that a blue vertex of Q represents a dihedral angle of an edge of P that is larger than its corresponding edge of P'.

Now we apply Cauchy's arm lemma (Lemma 6.36) extended to spherical polygons. Let chain C be the polygon Q with any edge e deleted. The lemma then applies to C and shows $|e'| > |e|$, where e' is the edge of Q' corresponding to e; see Figure 6.21(b). But we saw earlier that these lengths are fixed by the equal face angles incident to v from which they derive. Thus we arrive at a contradiction if there are no sign changes about v.

2. *There are exactly two sign changes.* In this situation, we can draw a chord s across the polygon Q separating the blue and red colors, as given in Figure 6.22(a). This cuts the polygon into two chains C^B and C^R, having blue and red colors, respectively. Applying Cauchy's arm lemma to C^B delimited by s shows that s must increase in length to become s' in Q': $|s'| > |s|$. Figure 6.22(b) shows the situation. Switching viewpoint to the red chain C^R shows that s must decrease in length to become s': $|s| > |s'|$. We again have reached a contradiction.

This completes the proof of Cauchy's rigidity theorem.   □

Exercise 6.41. *Prove Lemma 6.1: a polyhedron is convex if and only if all of its edges are convex. Use the spherical polygon viewpoint as depicted in Figure 6.20.*

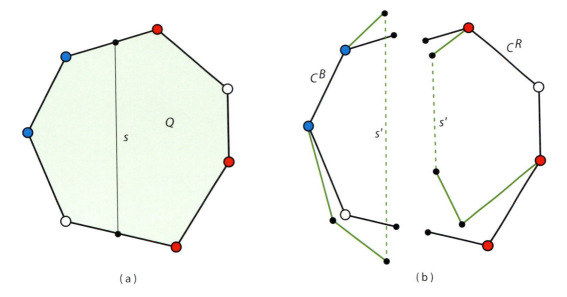

Figure 6.22. (a) The spherical polygon Q having exactly two sign changes. The chord s divides the polygon into sides having vertices of one kind of color. (b) The chains C^B and C^R show differing edge lengths of s', resulting in a contradiction.

Exercise 6.42. *Prove Lemma 6.3: for any convex polyhedron, the sum of the face angles incident to each vertex is at most 2π. Use the spherical polygon viewpoint.*

As is typical with a deep result like Cauchy's rigidity theorem, there have been many subsequent generalizations and developments. Alexander Alexandrov, a Ph.D. student of Delaunay, extended the uniqueness claim in 1948 in a surprising way. We may rephrase Cauchy's theorem as follows: If you glue together a collection of flat, rigid, polygonal faces so that every vertex has non-negative Gaussian curvature (i.e., such that the total angle at each vertex is $\leq 2\pi$), then the result is a unique convex polyhedron. In other words, the polygonal faces and their gluing uniquely determines the resulting convex polyhedron.

Alexandrov proved that if you glue together a collection of *flexible* flat polygons, again so that every vertex of the resulting closed surface has non-negative curvature, the result is still a *uniquely* determined convex polyhedron! The key aspect of this theorem is that, unlike in Cauchy's theorem, the edges of the polyhedron have no a priori relation to the edges of the polygons that are glued together. One remarkable consequence is that if one simply glues one half of the perimeter of a single convex polygon (perhaps made from paper) to the other half, the result is a unique convex polyhedron.

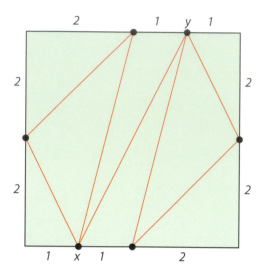

Figure 6.23. Gluing one half of the perimeter to the other half, creasing along the lines shown, results in an octahedron.

Exercise 6.43. *On a 4 × 4 square piece of paper, mark points x and y as shown in Figure 6.23. Their symmetric placement ensures that they partition the perimeter into two equal-length portions. Now glue the perimeter halves to one another, pinching at x and y, using tape to hold the identified segments to one another. With some coaxing, you should find that the paper wants to crease along the red lines shown, resulting in the unique octahedron which Alexandrov's theorem guarantees exists.*

We have thus far been discussing *rigidity*, but what about *flexibility*? How much freedom do we have on the periphery of Cauchy's theorem? If we remove the *closed* condition of the theorem, that is, if the polyhedron is permitted to have a boundary, the theorem fails. Cutting out a single polygonal hole that includes a (true, nonflat) vertex from the surface of a convex polyhedron necessarily renders it flexible. Figure 6.24 shows the flexing possible when an octahedron is cut through the middle. Note that such a hole leads to $V + E + F = 1$, so the reliance of Cauchy's proof on Euler characteristic 2 means that it no longer necessarily holds. A precise characterization of the conditions on the resulting polyhedral surface with boundary that distinguish rigidity from flexibility was worked out in 1958 by Alexandrov and L. A. Shor.

What happens if we remove the *convex* condition of the theorem? As we saw in Figure 6.16, if the polyhedron is not convex, the theorem fails. But this failure is demonstrated by two polyhedra which cannot be made into each other by a *continuous* motion. In other words,

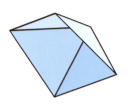

Figure 6.24. The flexing of half of an octahedron.

both of these polyhedra are *infinitesimally rigid* since they cannot be continuously deformed into any other configuration without tearing or bending. Cauchy's theorem guarantees that if a polyhedron is convex, it is infinitesimally rigid. But the question remained, if a polyhedron is nonconvex, can it *flex*? In a remarkable result, Herman Gluck proved in 1975 that *almost all* triangulated polyhedra are infinitesimally rigid, where "almost all" means that the rigid polyhedra form an open dense set in the space of all polyhedra. At this point in time, it was tantalizingly feasible that all triangulated polyhedra (homeomorphic to spheres), convex or nonconvex, are rigid.

In 1978, a few years after Gluck's result, Robert Connelly stunned the community by constructing a flexible polyhedron. Modifying a self-intersecting flexible octahedron which Raoul Bricard (a French engineer) had constructed in 1897, Connelly found the first example of a true flexible polyhedron, consisting of 18 triangular faces. Subsequent simplifications led to a 14-triangle, 9-vertex flexible nonconvex polyhedron constructed by Klaus Steffen. Figure 6.25 shows a construction by Rohan Mehra and Norman Nicolson of the Steffen polyhedron using plexiglass and piano hinges.

The existence of flexible polyhedra leads to a natural question: as the polyhedron flexes, does its volume change? For the 2D version, if we flex the vertices of a square, deforming it into a rhombus, clearly the areas of the polygons change, and it is natural to suppose the same holds in 3D. The surprise answer was proved by Idzhad Sabitov in 1997:

Theorem 6.44. *The volume of a flexible polyhedron does not change as it flexes.*

Thus a flexible polyhedron cannot serve as a bellows! Sabitov constructed a polynomial formula for the volume of any polyhedron, a formula that is based only on the combinatorial structure of the surface and the edge lengths. Because neither this structure nor the edge lengths change during a flex (only dihedral angles change), the volume must be constant. We close this section with a conjecture by Connelly.

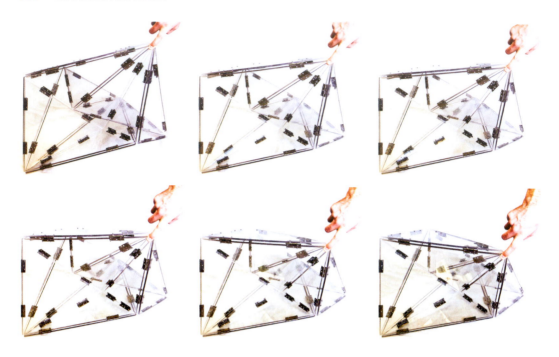

Figure 6.25. A construction of the Steffen flexible polyhedron using plexiglass.

UNSOLVED PROBLEM 23   Flexible Polyhedra

Show that the Dehn invariant (Section 1.5) of a flexible polyhedron does not change under flexing.

6.5 SHORTEST PATHS

A natural method to build a model of a polyhedron is to start with a connected planar layout of its faces which is then creased and folded to the polyhedron. Such a planar layout is called a *net* for the polyhedron. For example, one of the many possible nets for the truncated icosahedron of Figure 6.3(a) is shown in Figure 6.26.

The boundary of a net forms a polygon, and each face of the polyhedron appears whole inside the net. Reversing the viewpoint, imagine starting with a paper polyhedron, cutting the surface and unfolding to the plane. The unfolding produces a net if the cuts are along edges of the polyhedron, and the surface unfolds to a single piece without overlap.

Not every polyhedron has a net. Figure 6.27 demonstrates two examples, although proving that these nonconvex polyhedra have no nets is by no means straightforward. The polyhedron in (a) has six faces that are congruent nonconvex polygons; the extra edges needed to partition

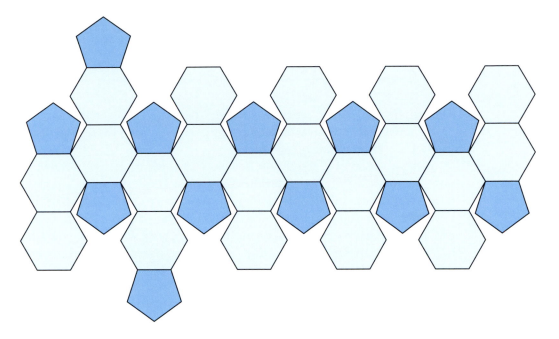

Figure 6.26. A net of the truncated icosahedron given in Figure 6.3(a).

these faces into convex pieces lead to a net. The nonconvex polyhedron in (b) also has no net, but here all of its 36 faces are just triangles. Both of our "netless" examples are nonconvex, and indeed every polyhedron proved to date to have no net is nonconvex. And yet this attractive problem remains open:

(a)   (b)

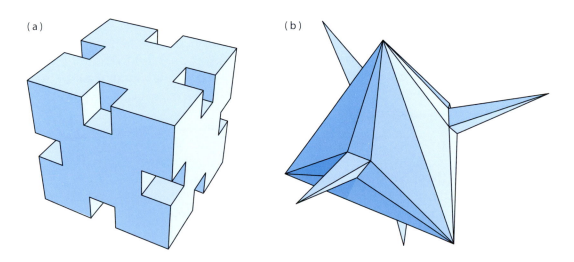

Figure 6.27. Nonconvex polyhedra without nets.

> ## UNSOLVED PROBLEM 24 Dürer's Problem
>
> Does every convex polyhedron have a net?

The problem is called "Dürer's problem" because Albrecht Dürer drew many nets for convex polyhedra as a way of presenting those polyhedra in his 1525 masterwork on geometry.

In the absence of a resolution of Dürer's problem, it is of interest to unfold convex polyhedra under different rules: allow the surface cuts to be arbitrary (rather than only along polyhedron edges), but still insist that the unfolding result in one non-overlapping piece, so that it could be cut out and refolded to the polyhedron. In such a *general net*, a polyhedron face may not appear whole; rather, it could be partitioned by cuts and distributed in the net, but in such as way that folding brings the pieces back together.

There are several methods known to produce a general net for any convex polyhedron: this freedom to cut anywhere is enough to resolve Dürer's problem. We will present two of these methods in this section. Both methods rely on shortest paths on the surface, a topic of considerable interest and many applications in its own right, aside from its relevance to unfolding. Shortest paths lead to beautiful connections to Voronoi diagrams and the medial axis. Shortest paths are also a subclass of geodesics, which we explore in Section 6.6 to close the chapter.

Exercise 6.45. *Find a net for each polyhedron in Figure 6.5.*

Exercise 6.46. *Prove that the polyhedron in Figure 6.27(a) has no net.*

Exercise 6.47. *Is there any edge unfolding of a cube that results in overlap?*

Exercise 6.48. *Construct a tetrahedron and an unfolding of it that self-overlaps.*

Even though all the material we present holds for more general surfaces, we restrict attention to a convex polyhedron P, and we insist that all of its vertices be "true" corners with positive curvature.

Definition. A *shortest path* on P between two points x and y on P is a curve connecting x and y whose length, measured on the surface, is shortest among all curves connecting those points on P.

There is always a shortest path between any two points, but it may not be unique: several distinct but equally shortest curves may connect the points. These equally short paths will play an important role below. Three fundamental properties of shortest paths we will need in the sequel are:

1. Shortest paths are simple in that they never self-cross.
2. A shortest path never passes through a vertex, although it may begin or end at a vertex.
3. If a shortest path σ passes through an interior point of an edge e, the planar unfolding of the two faces sharing e unfolds the two segments of σ on the faces to a single straight segment.

We leave proofs to the reader.

Exercise 6.49. *Prove that a shortest path never self-crosses.*

Exercise 6.50. *Prove that a shortest path never passes through a vertex.*

Exercise 6.51. *Prove that the planar unfolding of a shortest path is a straight line segment.*

A central goal of the earliest works on shortest paths in the computational geometry literature was to compute the shortest path on P from some fixed point x in P to any other point y in P. A more specific problem — computing the shortest path from x to all vertices of P — turns out to be just as difficult but easier to understand, so we start with this.

Example 6.52. Let P be the $2 \times 1 \times 1$ box with x the midpoint of the bottom face. The eight shortest paths to the vertices are shown in Figure 6.28(a). Property 3 above was used to compute the shortest paths to the vertices on the top face: unfolding the back and front faces of the box, as shown in part (b), allows the paths to become straight segments of length $\sqrt{1^2 + (3/2)^2} = \sqrt{13}/2 \approx 1.8$. Indeed, a path that first goes to a bottom vertex and then travels along a vertical edge of the box is considerably longer, with length $\sqrt{1^2 + (1/2)^2} + 1 = \sqrt{5}/2 + 1 \approx 2.1$.

The symmetry and simplicity of this example makes it seem that the task might be easy, but the goal appears more challenging when considering a more substantive example like that in Figure 6.29. When paths from x to all points are considered, the potential complications are more evident.

Recall that the cut locus (introduced by Poincaré and discussed in Section 5.1) marks where shortest paths are "cut" or terminated. For

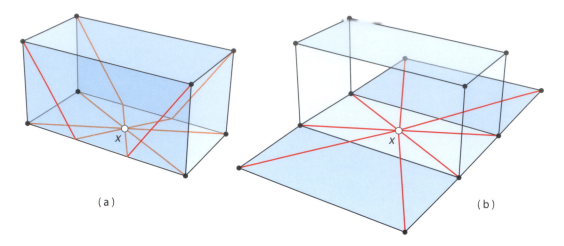

Figure 6.28. (a) Shortest paths from x to the eight vertices of a $2 \times 1 \times 1$ box. (b) Unfolding the back and front faces shows the shortest path is indeed a straight line.

our purposes, this is exactly the construct we need to identify the shortest path from x to an arbitrary point y.

Definition. The *cut locus* $\mathcal{C}(x)$ of x is the closure of the set of all points y to which there is more than one shortest path from x.

This definition should feel familiar because it is almost the same as the definition of the medial axis in Chapter 5. For the medial axis, the shortest paths are measured to the boundary of the polygon; here they

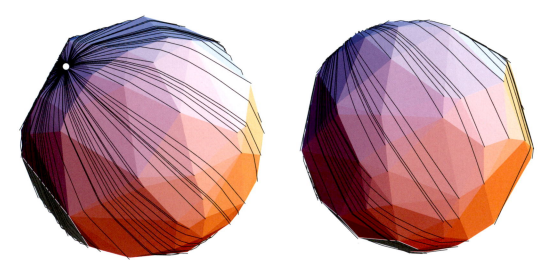

Figure 6.29. Two views of the shortest paths from a source point to all 100 vertices of a convex polyhedron. The shortest paths are lifted slightly from the surface to enhance visibility.

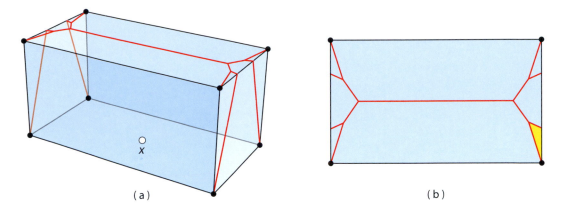

Figure 6.30. (a) The cut locus $\mathcal{C}(x)$ along with (b) a close-up of the top of the box.

are measured to a *source point x*. Many of the properties of the medial axis carry over in this new context.

Example 6.53. The cut locus of x for the $2 \times 1 \times 1$ box example is shown in Figure 6.30(a). It includes the central portion of the midline of the top of the box, evidently having two distinct shortest paths from x: up the front face, or symmetrically up the back face. However, near the left and right ends of the top, its structure becomes more complex. Part (b) shows a close-up of the top face. Here, for instance, the shortest path from x to any point in the shaded region follows the bottom-front-right-top faces. The resemblance of the cut locus to a Voronoi diagram will be explained later.

Exercise 6.54. *Sketch the cut locus $\mathcal{C}(x)$ for each of the five Platonic solids, where the source x is a vertex of the polyhedron.*

Assuming we knew $\mathcal{C}(x)$, then finding the shortest path to a point y in P would be easy. Identify the face f containing y and then locate y within a region of f determined by the partition induced by $\mathcal{C}(x)$. Finally, unfold the sequence of faces that reach that region, and connect x to y by a straight segment in the unfolding (due to Property 3). Similarly, knowing $\mathcal{C}(x)$ would suffice to find the shortest path to all vertices, and indeed there is essentially no easier way to find the paths to all vertices. So the focus of algorithms is to construct $\mathcal{C}(x)$. The following is a key property of $\mathcal{C}(x)$:

Theorem 6.55. *For any point x of a convex polyhedron P, the cut locus $\mathcal{C}(x)$ is a tree whose edges are straight segments on each face and whose leaves are the vertices of P.*

We will not attempt to prove this, but just point out again the similarity to the medial axis, which is also a tree terminating at vertices. Recall that a vertex v of P may or may not have two or more distinct shortest paths from x. In the box example, the paths to all vertices are unique. This is the reason for defining $\mathcal{C}(x)$ as the closure: in a neighborhood of a vertex v, there are points with two distinct shortest paths from x. The closure incorporates v into $\mathcal{C}(x)$ and therefore makes $\mathcal{C}(x)$ a tree spanning the vertices.

Exercise 6.56. *Argue that $\mathcal{C}(x)$ can have a cycle when P has genus greater than zero.*

Let's now consider algorithms for shortest paths, where we begin with an algorithm for computing $\mathcal{C}(x)$. We restrict ourselves to a high-level exposition because the details are formidable. The algorithm is a continuous version of a discrete algorithm for finding shortest paths in a graph, which we sketch first.

One of the earliest (1959) and still among the prettiest algorithms is Edsger Dijkstra's graph algorithm for finding shortest paths from a fixed source node x to every other node y in graph G, where distance is measured by non-negative weights assigned to each edge. Although the algorithm is general, for the purposes of illustration, we will specialize it to a plane graph G with the edge weight for each edge e given by its Euclidean length $|e|$. The algorithm can be viewed as a discrete simulation of the following continuous process. Imagine pouring paint on the source node x, and suppose the edges of G are thin pipes of the same diameter, so that the paint spreads evenly along all edges at a uniform rate, one unit of length per unit of time. Figure 6.31 shows how the paint spreads methodically until all vertices are reached.

Dijkstra's algorithm avoids a continuous simulation of the paint creeping down each edge, recognizing that discrete steps suffice. At all times t, the algorithm maintains a tree T rooted at x that spans all those vertices reached by the paint: the discrete *frontier*. At each step, the edges incident to every node of T are examined and an edge e is added to T that (a) reaches a node z outside T such that (b) the distance to x from z is shortest among all such nodes, breaking ties arbitrarily. Condition (a) ensures that the addition of e avoids a cycle, and condition (b) ensures that z is the next node to be reached by paint.

When the algorithm terminates, T is a spanning tree of all vertices in G, with the property that the shortest path from x to y in G is the unique path from x to y in T. The algorithm can easily be implemented to run in $O(n^2)$ time for a graph of n vertices, and with more care in $O(n \log n)$ time for graphs with $O(n)$ edges, such as planar graphs. Once completed, each query for a shortest path to y can be answered in $O(n)$ time.

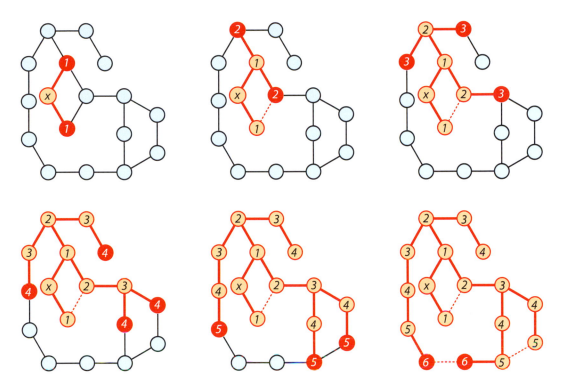

Figure 6.31. Dijkstra's algorithm on a plane graph G all of whose edges have unit length. The frontier nodes (shaded red) are labeled with the time at which they are reached. The tree T is shown at each step (excluding dashed edges).

The frontier at any stage of Dijkstra's algorithm is a discrete collection of graph nodes that have been reached from x up to some time t. Generalizing Dijkstra's algorithm to find shortest paths on the surface of a convex polyhedron P requires a *continuous* frontier, in fact, a collection of curves on the surface of P composed of circular arcs. Figure 6.32 illustrates a snapshot of such a frontier for the $2 \times 1 \times 1$ box from Example 6.53 at $t = 3/2$: the locus of points at distance $3/2$ from x. The two "wavefronts" just touching the top face at $t = 3/2$ will meet at the midpoint of that face at $t = 2$, and then proceed to intersect along the midline for $t > 2$. Note that the junction between two wavefront arcs lies either on an edge of P or on the cut locus $\mathcal{C}(x)$. Tracking this frontier then implicitly computes $\mathcal{C}(x)$.

This approach to computing shortest paths on a polyhedron was first explored in an influential 1987 paper by Joseph Mitchell, David Mount, and Christos Papadimitriou. The frontier curves are maintained implicitly and the algorithm still advances in discrete spurts. Rather than the event being the frontier reaching a new node, the key event is now the frontier reaching an edge from one side. But because several distinct wavefronts can reach one edge, from either side, edges must be partitioned into intervals "owned" by particular bundles of shortest paths, distinguished

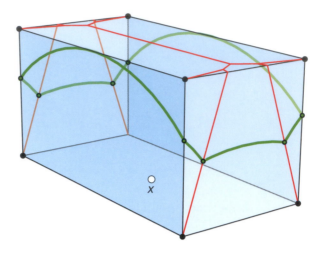

Figure 6.32. A continuous frontier on the 2 × 1 × 1 box.

by the combinatorial sequence of edges they cross. The final result is an $O(n^2 \log n)$-time algorithm which, remarkably, works for nonconvex polyhedra as well.

The two decades since 1987 saw a long pursuit by the community for an algorithm for convex polyhedra that matched the time complexity of Dijkstra's algorithm. An advance to $O(n^2)$ time was achieved after a decade, which, despite the small improvement, was an eminently implementable algorithm. In fact, this algorithm was used to produce Figure 6.29. Another decade finally saw in 2008 an $O(n \log n)$-time algorithm, still following the continuous Dijkstra approach but with intricate and delicate data structures, detailed in an 80-page paper by Yevgeny Schreiber and Micha Sharir.

Now that we have described how to compute $\mathcal{C}(x)$, we return to the topic of unfolding to *general* nets, where the cuts are not restricted to be along edges. The two methods we mentioned at the beginning of this section for producing a general net for any convex polyhedron are known as the source unfolding and the star unfolding. The latter is easiest to describe, so we start with this method.

Definition. The *star unfolding* of a convex polyhedron P fixes a generic[4] point x on P, finds the shortest paths from x to every vertex of P, and simply cuts along these shortest paths.

---

[4] The source point x is generic when there is a unique shortest path to each vertex, as in Figure 6.28.

If P has n vertices, then the star unfolding $U^*(P)$ is a polygon of $2n$ vertices: n vertices from the polyhedron and n vertices from n copies of x. The reason these cuts suffice to flatten is that all the points of curvature are "resolved" by these cuts. Indeed, it is a necessary condition for flattening to a general net that the cuts form a spanning tree of the vertices — spanning in order to reach all points of positive curvature, and a tree so that the unfolding is one piece. (A cut cycle would isolate a separate piece of the surface.)

Example 6.57. Figure 6.33 shows the star unfolding produced by cutting the eight paths in Figure 6.28(a). Underneath are shown the box faces from which pieces of the star unfolding derive, arrayed about the top face affixed to the plane. Consider the vertices of $U^*(P)$: The eight copies of the source point x are interleaved with the eight copies of the vertices of P. Notice that the two polygon edges incident to a copy v' of a vertex v of P have the same length — they are the two "sides" of the cut along the shortest path from x to v on P. Moreover, the exterior angle of $U^*(P)$ at v' is the curvature of v on P.

That $U^*(P)$ does not self-overlap, and so is in fact a simple planar polygon, is by no means obvious. The notion of a star unfolding goes

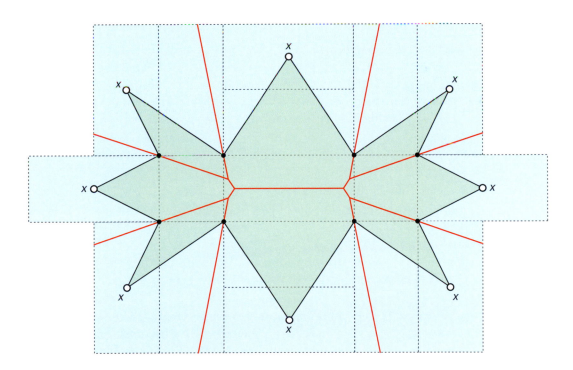

Figure 6.33. The star unfolding of the box in Figure 6.28(a). The red lines form the Voronoi diagram of the eight copies of the source point x. Notice the cut locus $\mathcal{C}(x)$ appears as a subset of this Voronoi diagram.

back to Alexadrov in 1948 but it was only shown to avoid overlap in 1992. This and the relationship to Voronoi diagrams are gathered in the following theorem:

Theorem 6.58 (Star Unfolding). *The star unfolding from a generic point x on a convex polyhedron P with n vertices is a simple (nonoverlapping) polygon $U^*(P)$ with $2n$ vertices. Furthermore, the Voronoi diagram of the n copies of the source x, restricted to the interior of $U^*(P)$, is the unfolding of the cut locus $C(x)$ on P.*

Thus the cut locus *is* a Voronoi diagram. The single source point x has as many unfolded images of x as there are vertices of P, and the cut locus $C(x)$ is the Voronoi diagram of these images. In general, this Voronoi diagram is not a tree, but when clipped to the interior of $U^*(P)$, it leaves exactly $C(x)$. Note that $C(x)$ is indeed incident to each vertex v' of $U^*(P)$ because it spans the vertices of P.

Although one might think this theorem gives a new route to computing the cut locus, in fact the algorithms that find the shortest paths to all vertices effectively compute the cut locus along the way, as we saw with the continuous Dijkstra approach. Nevertheless, the insight the theorem gives into the structure of the cut locus is useful, and the relationship has found several algorithmic applications.

Exercise 6.59. *Sketch the star unfolding of each of the Platonic solids with the source x at a vertex, breaking shortest-path ties arbitrarily, so that there are $n - 1$ paths from x for a polyhedron of n vertices. (A vertex is definitely not generic, but the star unfolding still avoids overlap.) The dodecahedron and the icosahedron are the most difficult.*

Now we turn to another unfolding of a convex polyhedron to a general net.

Definition. The *source unfolding* of a convex polyhedron P is simply obtained by cutting the cut locus $C(x)$ of a generic source point x.

Recall that the cut locus is a spanning tree of the vertices of P, which we know is a necessary condition for flattening. Let $U^s(P)$ be the flattened source unfolding for polyhedron P. Why is $U^s(P)$ a simple nonoverlapping polygon? In contrast to the star unfolding nonoverlap proof, this is easy: Consider Figure 6.34 which shows the unfolding produced by cutting $C(x)$ in Figure 6.30(a). The source x is at the center of the polygon $U^s(P)$, with $C(x)$ forming its boundary. The shortest paths

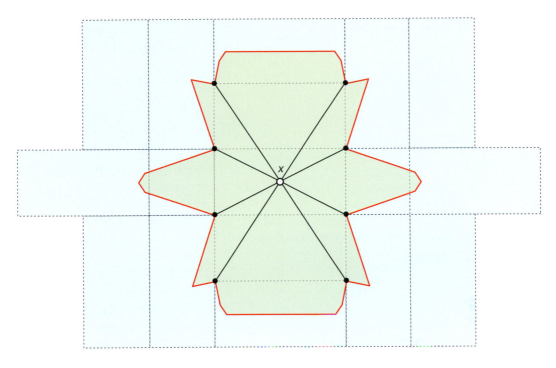

Figure 6.34. The source unfolding of the box in Figure 6.30(a), with the cut locus (in red) forming its boundary. The shortest paths from x to the eight vertices of the box are drawn.

(sprayed the full $360°$ around x) fill the surface of P and so fill $U^s(P)$. None of these shortest paths cross, and so the source unfolding is a simple polygon, visible from x.

Because the source and the star unfoldings represent the same surface P, they are scissors congruent, in the terminology of Section 1.4. Indeed, comparing Figures 6.33 and 6.34, one can see that the cut locus partitions the star unfolding into convex polygons that may be reassembled to form the source unfolding.

Exercise 6.60. *Show that cutting $C(x)$ from a nongeneric point x of P might not lead to a spanning tree on P, and so would not permit flattening to a plane.*

★ **Exercise 6.61.** *Find a polyhedron P and a point x so that $U^*(x)$ is congruent to $U^s(x)$: the source and the star unfoldings are congruent polygons.*

Now that we have seen every *convex* polyhedron having a general net, it is natural to wonder whether every polyhedron, convex or nonconvex, has a general net. This question remains temptingly open:

> **UNSOLVED PROBLEM 25**   General Nets
>
> Does every polyhedron have a general net?

6.6 GEODESICS

We close this chapter with an old and deep theorem of Lyusternik and Schnirelmann: every closed surface has at least three distinct, simple closed geodesics. Even though the theorem applies to surfaces homeomorphic to a sphere, we will specialize to convex polyhedra, where, fortunately, the theorem loses none of its force or charm. We first begin with a definition.

Definition. A *geodesic* on a surface is a curve γ with the property that for any two sufficiently close points x and y on γ, the portion of γ between x and y is the shortest path on the surface connecting x and y.

Thus geodesics are locally shortest paths, that is, locally length minimizing. Every shortest path is a geodesic, but geodesics are often not shortest paths. Although shortest paths terminate at the cut locus, geodesics continue beyond. And as they continue, they can self-cross, a phenomenon not possible for a shortest path. Figure 6.35(a) shows part of a geodesic on a cube that self-crosses.

A geodesic that does not cross itself is called a *simple geodesic*. If in addition the geodesic forms a closed loop on the polyhedron, joining smoothly to itself, it is called a *simple closed geodesic*. Figure 6.35(b) provides such an example on the cube. It was conjectured after investigations by Poincaré that there must be at least three simple closed geodesics

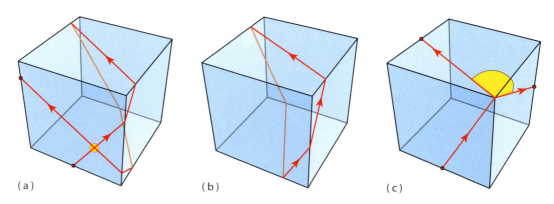

(a)   (b)   (c)

Figure 6.35. (a) Part of a self-crossing geodesic on a cube. (b) A simple closed geodesic. (c) A quasigeodesic can take many paths through a vertex.

on any surface homeomorphic to a sphere, and a proof was offered by Lazar Lyusternik and Lev Schnirelmann in 1929 in a two-page paper. Their technique used Birkhoff shortening, but it is generally agreed that this plan was not firmly established until recently. So it may be that Matthew Grayson's employment of curve shortening from 1989 was the first technically correct proof of the Lyusternik-Schnirelmann theorem. It is this approach we highlight here, but in the context of polyhedral surfaces.

In general, a convex polyhedron does not admit any simple closed geodesic. This is easily seen: because a geodesic does not turn on the surface, its total turn $\tau = 0$ and the Gauss-Bonnet theorem, in the form of equation (6.8), says that the total curvature on each side of the geodesic must be 2π.[5] But, because shortest paths do not pass through vertices (Property 2) and geodesics are locally shortest, geodesics do not pass through vertices either. Thus, with curvature being concentrated at vertices, there must be a partition of the vertices into exactly 2π curvature in each half. But this would occur with probability zero for a generic polyhedron! Note the closed geodesic in Figure 6.35(b) indeed does partition the total curvature equally, as it has four vertices to either side.

There is a natural extension that retrieves the Lyusternik-Schnirelmann theorem for convex polyhedra through the notion of *quasigeodesics*, as introduced by Alexandrov and further developed by his student Aleksei Pogorelov:

Definition. There are two angles defined at each point x on a directed curve on a polyhedron: $R(x)$ is the total incident face angle at x to the right at x, and $L(x)$ the angle to the left. A *quasigeodesic* is a curve γ on a surface such that both $R(x) \leq \pi$ and $L(x) \leq \pi$ at every point x in γ.

A geodesic is a quasigeodesic at every point x not coincident with a polyhedron vertex because $R(x) = L(x) = \pi$. This is true even on interior points of edges. When a geodesic passes through a polyhedron vertex, however, there are many continuations that could be unfolded straight. For example, in Figure 6.35(c), the geodesic that hits the cube corner could emerge anywhere within the indicated $\pi/2$ range of angles and have at most π angle to either side. So Alexandrov's definition is indeed the natural definition from the point of view of unfolding to a straight line: quasigeodesics are exactly those curves that can unfold to straight lines, as per Property 3 of shortest paths.

---

[5] In fact, Poincaré planned to prove that the shortest closed curve dividing the 4π curvature equally must be a simple closed geodesic, a plan that waited until 1982 to be completed by Christopher Croke.

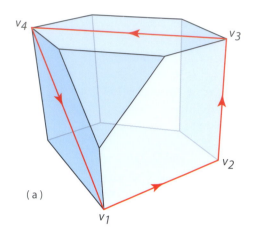

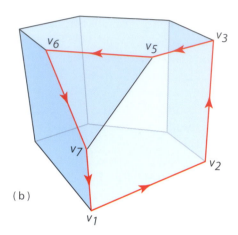

Figure 6.36. (a) A simple closed quasigeodesic marked on a convex polyhedron; (b) another closed curve that is not a quasigeodesic.

Theorem 6.62 (Lyusternik-Schnirelmann-Pogorelov). *Every convex polyhedron has at least three distinct simple closed quasigeodesics.*

Example 6.63. Figure 6.36(a) shows a simple closed quasigeodesic γ on a truncated cube passing through four vertices (v_1, v_2, v_3, v_4). The angles to either side at the vertices v_i of γ are, respectively,

$$R(v_i) : \left(\frac{3}{4}\pi, \pi, \frac{3}{4}\pi, \pi\right),$$

$$L(v_i) : \left(\frac{3}{4}\pi, \frac{1}{2}\pi, \frac{3}{4}\pi, \frac{1}{2}\pi\right).$$

So γ is indeed a quasigeodesic. In contrast, the curve shown in Figure 6.36(b) is not a quasigeodesic because

$$L(v_5) = L(v_7) = \frac{3}{4}\pi + \frac{1}{3}\pi > \pi.$$

Although simple closed quasigeodesics are known to exist, there is effectively no way to find them!

> **UNSOLVED PROBLEM 26**   Finding Quasigeodesics
> Find a polynomial-time algorithm for constructing even one of the three quasigeodesics.

Exercise 6.64. *Argue that the boundary of a face of a regular tetrahedron is a simple closed quasigeodesic.*

Exercise 6.65. *Find two more simple closed quasigeodesics for P in Figure 6.36. In this example, each vertex deriving from a truncation is a midpoint of the cube edge on which it lies. Thus the front face is an equilateral triangle and the back faces are rectangles.*

★ **Exercise 6.66.** *Find all the simple closed geodesics on the regular tetrahedron.*

★ **Exercise 6.67.** *Find all the tetrahedra that admit simple quasigeodesic infinite lines of unbounded length.*

We close this chapter by returning to the case of a smooth convex surface. By Lyusternik-Schnirelmann, the surface contains at least three simple closed geodesics. Although no efficient method is known for finding the three geodesics, it is possible by using a simulation of the curve-shortening proof of Gage, Grayson, and Hamilton, which we now sketch. Let S be a smooth convex surface and imagine slicing it with parallel planes, producing an infinite stack of closed curves, and two points where the planes are just tangent to S; see Figure 6.37 for the idea.

Figure 6.37. Parallel slices through a convex body.

The 1-parameter family of curves constitutes a cover V of S. Applying curve shortening to each curve in V continuously deforms the cover, but retains it as a cover for S. Let γ be a particular curve in V. Grayson

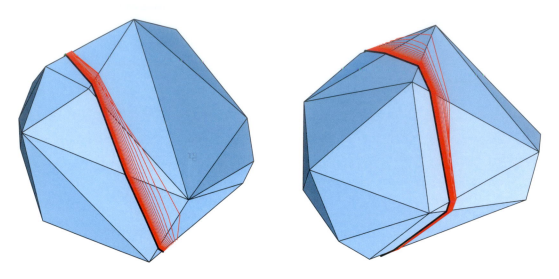

Figure 6.38. Two views of a polyhedron showing simulations of curve shortening to a simple closed quasigeodesic.

proved that either γ evolves under the curve-shortening flow to a round point, effectively slipping off S, or γ "exists for all time," in which case its curvature converges to zero, that is, it becomes a geodesic.

Now we know the Gauss-Bonnet theorem holds for S and so S has a total of 4π of curvature. Moreover, some curve γ_0 in the initial slicing must split the curvature into $2\pi + 2\pi$. A property of the curve-shortening flow is that the evolution of γ_0 must preserve this curvature-partitioning property. Thus γ_0 cannot converge to a round point (because a round point cannot contain 2π of curvature), and so it must converge to a geodesic. This establishes that there is at least one simple closed geodesic on S. That there are *three* distinct such geodesics follows from a topological argument involving homology classes of curves that we will not attempt to detail.

Although we have only sketched the argument for smooth convex surfaces, the result holds for any smooth surface homeomorphic to a sphere. Furthermore, the above proof can be mimicked to numerically find a simple closed quasigeodesic on convex polyhedra, as illustrated in Figure 6.38. Whether this approach can be proved to always converge, and in polynomial time, and thus resolve the unsolved problem posed above, is unknown.

SUGGESTED READINGS

Peter Cromwell. *Polyhedra*. Cambridge University Press, 1997.
> A delightfully readable source on the history of polyhedra, including clear proofs of Euler's theorem and Cauchy's rigidity theorem.

H. S. M. Coxeter. *Regular Polytopes*. Dover, 2nd edition, 1973.
> Detailed analysis of regular polytopes in four and higher dimensions by one of the originators of the field. The source of Coxeter groups, Coxeter graphs, and other concepts now named after him.

Branko Grünbaum. *Convex Polytopes*. Springer-Verlag, 2nd edition, 2003.
> A comprehensive and highly influential work on polytopes. It was originally published in 1967 and inspired a generation of researchers. The recently updated and expanded edition serves as a wonderful reference today.

Martin Aigner and Günter Ziegler. *Proofs from* THE BOOK. Springer-Verlag, 4th edition, 2009.
> Paul Erdős imagined that God has a book containing the best proof of every theorem. So when someone discovers a particularly pretty proof, the joke is that it is from THE BOOK. We drew on their clear exposition of Cauchy's rigidity theorem.

Alexander Alexandrov. *Convex Polyhedra*. Springer-Verlag, 2005.
> This seminal work from 1950 was not translated into English until 2005. It contains a full proof of Alexandrov's powerful extension of Cauchy's theorem, material on the flexibility of polyhedra, unbounded polyhedra, and many other topics.

Erik Demaine and Joseph O'Rourke. *Geometric Folding Algorithms: Linkages, Origami, Polyhedra*. Cambridge University Press, 2007.
> Many of the topics in the chapter are covered in greater depth in this monograph, especially the material on rigidity, unfolding, and quasigeodesics.

7 CONFIGURATION SPACES

We begin this chapter by revisiting translational motion planning using the Minkowski sum from Section 5.3, but then turn to more complex spaces for rigid objects (Section 7.1). Next we explore configuration spaces for the simplest articulated objects, the open polygonal chains (Section 7.2), which brings us to several famous problems on locked chains (Section 7.3). This in turn leads to an investigation of closed polygonal chains, concentrating on the topology of the space of polygons (Section 7.4). We close the chapter and the book with an advanced topic (Section 7.5), the structure of the configuration space of moving and colliding particles, leading back to the intricate combinatorics of the associahedron from Section 3.3.

7.1 MOTION PLANNING

The motion of objects through a perhaps cluttered environment can be understood by studying the space of all positions, or placements, or *configurations* of these objects. A *configuration space* C is a set whose elements consist of all possible configurations or arrangements of a set of objects. An example of a configuration space is the set of possible placements of chess pieces on a board that are realizable by legitimate moves in a chess game. The elements (or *points*) in the configuration space are the individual formations. Each chess move in any game corresponds to a simple connection between two points in the configuration space — the formation before the move and the formation after the move. In this particular example, the chess formations provide a *discrete* configuration space. Another classical example deals with robot motion planning, where the space of possible motions of robots inside a factory layout provides a *continuous* configuration space. Here each point in this space corresponds to a particular, distinct configuration of the robots.

The progression of the chapter is from more algorithmic concerns — whether a movement in configuration space between given positions is possible, and if so, how to realize the motion — to focus more on the topology and combinatorial structure of particular configuration spaces. A word on terminology. Although it is a crude simplification, physicists use *state space*, computer scientists use *configuration space*, and mathematicians use *moduli space* for the same concept. "State" emphasizes the status of a system determined by many perhaps disparate variables (such as temperature, pressure, and spin). "Configuration" suggests that geometric variables determine the object placement (such as translation coordinates and rotation angles). "Moduli" derives from "modulus," which in this context means some abstract mathematical parameter. We will use configuration space throughout, as our examples are geometric, but the alternative terms are prevalent in the literature.

We start our study of configuration spaces with a particularly simple *motion planning* problem: Find a path (if one exists) to move a single polygonal object R, without rotation, through the plane $\mathbb{R}^2$ littered with polygonal obstacles $P_1, P_2, \ldots, P_m$. One can think of R as a wheeled robot moving through a floor plan with chairs, tables, and walls to avoid. (Later, we will incorporate rotation after analyzing translational motion.) This is one instance of a vast array of problems which we will only sample. Although we increase the complexity of our examples throughout this chapter, we will by no means exhaust the full range of possibilities.[1]

We already indicated how to solve this translational motion planning problem in Section 5.3 on Minkowski sums, and in particular, in the example of Figure 5.12. We review and summarize the method here: Let s be the initial reference point of robot R and t the desired location of R, again specified by its reference point. At a high level, the algorithm is as follows:

MOTION PLANNING   Voronoi Diagram Algorithm $O(kn\log^2 n)$

Begin by growing all obstacles P_i via the Minkowski sum $P_i^+ = P_i \oplus -R$. Let P^+ be the union of the sums P_i^+ and consider its complement $\mathbb{R}^2 \setminus P^+$ in the plane. Find a path, if one exists, between s and t in this complement.

The left side of Figure 7.1 illustrates this algorithm for a quadrilateral robot R and eight polygonal obstacles. The complement of the Minkowski sum $\mathbb{R}^2 \setminus P^+$ is shown on the right (in white). This complement *is* the configuration space $\mathcal{C}$ of translations of robot R in the plane, where each point in this configuration space corresponds to a possible position of R. Notice that this space consists of three connected components A, B, C. Thus, from the initial position s (labeled in the figure), deciding whether a goal position t is reachable from s is reduced to determining whether s are t are in the same connected component of the configuration space. And planning a path for the robot reduces to finding a path for the reference point within this component.

---

[1] A recent 800-page book on motion planning contains more than 1000 references!

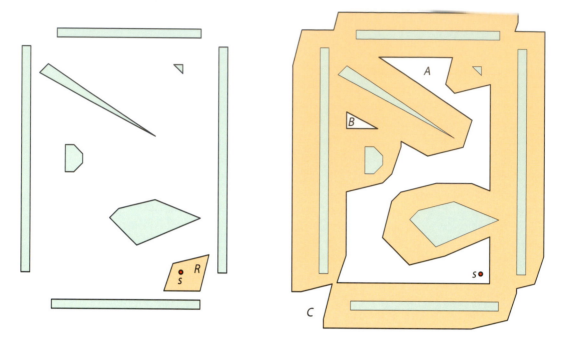

Figure 7.1. The left side shows robot R with reference point s along with polygonal obstacles. The right side shows the configuration space partitioned into three connected components A, B, C, deriving from complements of Minkowski sums with $-R$.

We have described the first step of this algorithm, computing the Minkowski sum, in some detail in Section 5.3. The other steps (forming unions, finding connected components, finding paths) present interesting algorithmic issues, none of which we will explore here. We choose instead to concentrate on the configuration space aspects. Suffice it to say that if R has k vertices and all the obstacles together have n vertices, then a path can be found in time slightly more than $O(kn \log^2 n)$, which is remarkably efficient.

Let's now consider issues with translation and rotation of objects. For translational motion of a rigid object in the plane, the configuration space $\mathcal{C}$ is a 2D space, where each point of $\mathcal{C}$ represents a placement or configuration of R determined by its reference point. Now when we allow R to rotate as well as to translate (still in the plane), $\mathcal{C}$ becomes a 3D space given by three *degrees of freedom* for its motion: translation by x, translation by y, and rotation by an angle θ (about the reference point of R). So each point (x, y, θ) of $\mathcal{C}$ represents an oriented placement of R within the 2D plane. The motion planning algorithm is identical in overall structure but matters are more complicated because $\mathcal{C}$ is more complicated.

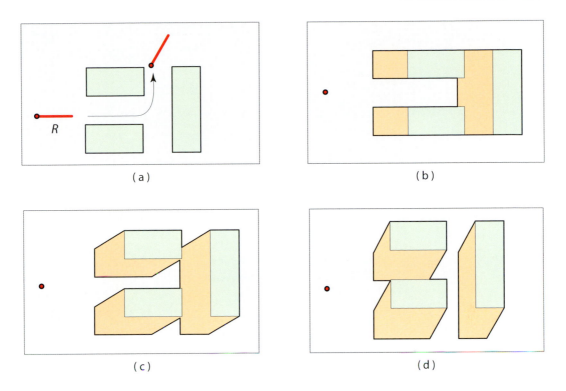

Figure 7.2. (a) The ladder R can reach the final position by rotating. The grown obstacles for (b) $\theta = 0°$, (c) $\theta = 30°$, and (d) $\theta = 60°$.

One way to understand what happens is to fix θ at successive values, and for each value, grow the obstacles by the Minkowski sum in 2D. Figure 7.2 illustrates this idea for the simplest possible extended object R, a line segment. Often this is called a *ladder* in the literature, reminding us of the challenge of carrying a ladder through a cluttered environment. As shown in part (a), the ladder can pass between the two horizontal polygons and then move up to the final position by rotating. If the reference point of R is the left endpoint, the grown obstacles are shown for (b) $\theta = 0°$, (c) $\theta = 30°$, and (d) $\theta = 60°$ by taking the Minkowski sum with $-R$. Now we imagine stacking the grown Minkowski sum obstacles along a third θ-axis. Considering the complement of these stacked obstacles in $\mathbb{R}^3$ yields the configuration space $\mathcal{C}$, the space of all possible ways the ladder can translate and rotate through the obstacles in the plane. Figure 7.3 shows two different views of part of the configuration space, as θ varies from $0°$ to $75°$. The stacking illustrated in this figure is discrete, but of course the configuration space varies continuously with θ.

As might be imagined, each step of this motion planning problem is more complicated than it was for translation-only. In particular, the grown obstacles are not polyhedral, as can be discerned in Figure 7.3: in each θ-plane they are polygonal, but they twist along the θ direction, producing shapes with curved edges bounding curved surface patches. Although it may have been justified above to leave to intuition that the

(a)

(b)

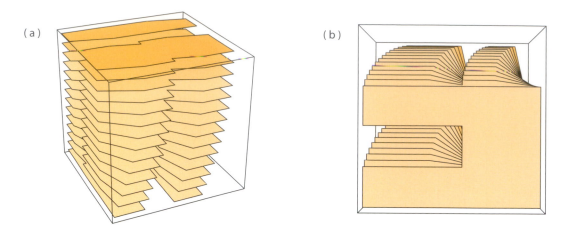

Figure 7.3. (a) The front view of the configuration space with θ varying from $0°$ to $75°$, and (b) a view from underneath the same stacks.

steps of the algorithm can be carried out when $\mathcal{C}$ is 2D and the grown obstacles are polygonal, in the situation considered here, where $\mathcal{C}$ is 3D, it is not clear how to implement any step of the algorithm: from the Minkowski sum, to the union, to finding a path.

Exercise 7.1. *How many dimensions does the configuration space have for moving a polyhedron in 3D, permitting both translation and rotation?*

We now sketch one general method, *cell decomposition*, which has the advantage of working for essentially all motion planning problems, even those much more complicated than we consider here. Cell decomposition is not only a completely general method, it was also the first developed. We describe it as applied to moving a ladder and only at the end discuss the generalizations. The essence of the cell decomposition approach is to partition the unruly configuration space into a finite number of well-behaved *cells*, and to determine a path in the space by finding a path between cells.

Consider the polygonal environment shown in Figure 7.4, consisting of two triangle obstacles and an open polygonal wall. The ladder R is allowed to translate and rotate in this region, where, as before, the reference point of the ladder is its left endpoint. To get a precise definition of a cell, we assign labels to every obstacle edge,[2] as is done in the figure; we use the label ∞ to represent a "surrounding" edge infinitely far to the right (just to reduce the number of labels in this example).

---

[2] Labels also should be assigned to the vertices, but we will ignore this minor complication here.

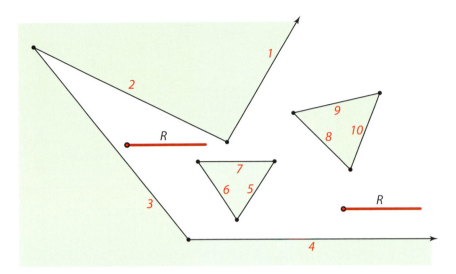

Figure 7.4. A polygonal environment for a ladder R, with obstacle edges labeled.

We begin by temporarily fixing the rotation of R to be $\theta = 0$. As always, the configuration space is what remains after growing the obstacles by the Minkowski sum with $-R$. Now suppose we place the ladder's reference point at a point x not on the same horizontal as any vertex of the polygonal environment. Then moving R horizontally leftward will cause it to eventually bump into an obstacle edge e_1, as will moving it horizontally rightward into edge e_2. Label the point x with this pair (e_1, e_2) of obstacle edges. A *cell* is a connected collection of points in the configuration space, all with the same label pairs. Figure 7.5(a) depicts the configuration space (the white regions) for a horizontal ladder along with its partition into eight labeled cells. Here no cell has a $(3, 6)$ label because such points do not exist in the configuration space; in other words, the ladder cannot fit between those two edges.

In the cell decomposition approach, the cell structure is represented by a graph, the *connectivity graph* G_θ. The subscript indicates that this graph captures the structure for a particular rotation of the ladder θ. The nodes of G_θ are the cells; two nodes are connected by an arc if the cells touch, or more precisely, if their boundaries share a nonzero-length segment. Indeed, G_θ is a type of dual graph to the partition of the configuration space by the cells.

The importance of this graph is that motion planning within a cell is trivial, so that a path in the graph can be easily converted into a path for the ladder. Moreover, the ladder can only move from one cell to another if there exists a path in the graph between these cells. Figure 7.5(b) shows the graph G_0 associated to part (a). Note that G_0 is disconnected: there is no path in G_0 between cells $(1, 8)$ and $(1, 9)$. Practically this means that

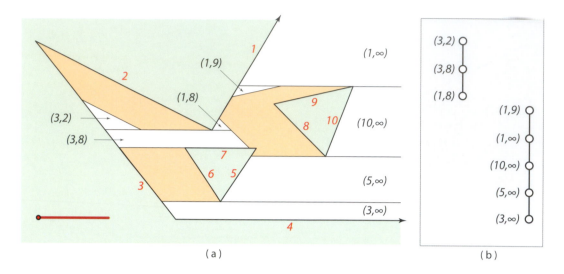

Figure 7.5. (a) The configuration space (white regions) for a horizontal ladder along with its labeled cell decomposition. (b) The connectivity graph G_0 corresponding to part (a).

one of the ladder positions shown in Figure 7.4 cannot be moved to the other position using horizontal movements only.

Now we incorporate rotation in a manner similar to the plane-stacking idea used previously. If we rotate the ladder slightly, the connectivity graph for the obstacles normally will not change: All the cells will change shape, but they will remain, and will maintain their cell adjacencies. But when the rotation exceeds some *critical rotation* θ^*, the combinatorial structure of G_{θ^*} will be different from that of G_0.

The horizontal $\theta = 0$ rotation shown in Figure 7.5(a) is critical because there are obstacle edges parallel to R (edges 4 and 7). Thus a slight rotation of R counterclockwise about its reference point will create a $(7, 8)$ cell, and a $(4, \infty)$ cell. Further rotation to $\theta = 13°$ (the angle of obstacle edge 9) causes cell $(1, 9)$ to disappear. The configuration space for $\theta = 13°$ along with its cell decomposition is shown in Figure 7.6(a); its corresponding connectivity graph is given in part (b) of the figure.

Exercise 7.2. *Create a ladder-moving example in which the cell decompositions of the configuration space labeled with some particular pair (i, j) are not all connected together.*

★ **Exercise 7.3.** *Suppose in the plane $\mathbb{R}^2$ there is a point obstacle at each integer lattice point. (We can think of the points as the base of poles sticking into $\mathbb{R}^3$.) What is the maximum length L of a ladder that can move from any one configuration to any another in this environment via planar translations and rotations? In other words, if the ladder endpoint at (x_1, y_1) and tilted at θ_1 is a valid position (it passes through*

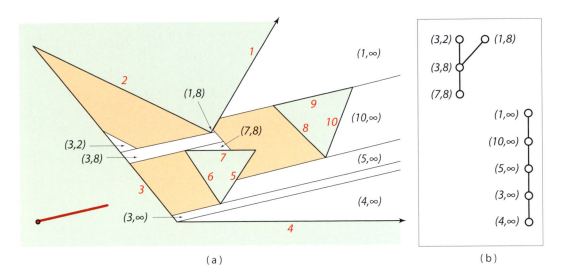

Figure 7.6. (a) The configuration space (white region) for a ladder rotated $\theta = 13°$, along with its labeled cell decomposition. (b) The connectivity graph G_θ corresponding to (a).

> *no obstacle points), and (x_2, y_2) and θ_2 is another valid position, then there should be a movement between these configurations.*

As should be clear now, critical rotation angles all involve the alignment of the ladder with either edges of obstacles, or two obstacle vertices. Thus there are at most an order of $O(n^2)$ critical angles. Now the idea is to form one grand connectivity graph G that incorporates the information in all the G_θ graphs. We extend the definition of a cell to represent regions of the 3D configuration space, all of whose points have the same forward/backward label pairs. This amounts to stacking the cells for fixed rotations on top of one another in the θ direction. Thus the points in cell $(3, 2)$ of Figure 7.5(a) are in the same 3D cell as the points in cell $(3, 2)$ of Figure 7.6(a). Each distinct 3D cell is a node of G, and again two nodes are connected by an arc if their cells touch, which now means that they share a nonzero-area boundary section. Figure 7.7 shows the graph G for all rotations of θ in $[0°, 90°]$ in our example. Note that the two positions illustrated in Figure 7.4 are indeed connected in G: cell $(3, 2)$ is connected to $(5, \infty)$.

The key to the construction of G is realizing that not all rotations of θ are needed — only the *critical* ones! The infinitely many layers of rotations are reduced to a finite number, and a continuous space (such as Figure 7.3) is converted to a discrete graph (such as Figure 7.7). The graph G then may be constructed by building G_0 and moving through all critical rotations in sorted order, modifying G_θ along the way, and

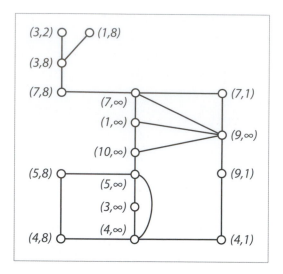

Figure 7.7. The graph G encompassing all rotations of the ladder from Figure 7.4 within the range $[0°, 90°]$.

incorporating the changes into G. We will not present any details, but at least now the construction of G should seem feasible. Again motion planning within a single cell represented by a node of G is not difficult, and moving between touching cells is also not difficult. For example, one could move from the interior of a cell to its boundary, and then move along the boundary to the portion shared with an adjacent cell. So the problem of motion planning is reduced to a graph problem: finding a path between the node corresponding to the cell containing s, to the node corresponding to the cell containing t. If there is no such path in G, then there is no path for the ladder, and if there is a path in G, it can be used as a guide for planning the motion of the ladder.

Careful implementation of the cell decomposition algorithm for translation and rotation achieves $O(n^2)$ time for moving a ladder, and slightly more than $O(k^2 n^2)$ time for moving a polygon with k vertices, where again n is the total number of obstacle vertices. As mentioned, the technique works in general, achieving a complexity doubly exponential, roughly $O(n^{3^d})$ time, where d is the dimension of the configuration space. This complexity was subsequently reduced by John Canny in 1987 with a different algorithm, called the *roadmap algorithm*, to singly exponential $O(n^{d+1})$ time (ignoring certain details). The roadmap algorithm reduces the multidimensional cells of the configuration space to a network of connections (the "roadmap") that suffice to move between any two points within the same component of $\mathcal{C}$. The reduction to the network employs a tool called "Whitney stratification," which would take us too far afield even to sketch. Suffice it to say that the roadmap algorithm, as subsequently developed, is the best general-purpose motion-planning algorithm. Although its exponential complexity is formidable, often the dimension d (the number of degrees of freedom) is a small constant. And with effort, the most important cases, such as those we have considered, have improved on the general algorithms.

Exercise 7.4. *Find a planar environment for moving a ladder such that C has $\Omega(n^2)$ connected components.*

Exercise 7.5. *What is the smallest doorway through which a convex polygon may pass? The doorway is a gap in an infinite line in the plane. The polygon may translate and rotate.*

★ **Exercise 7.6.** *Find a planar environment for moving a ladder such that the connected component that contains s and t requires the ladder to make $\Omega(n^2)$ moves, under any reasonable definition of what constitutes a distinct "move."*

7.2 POLYGONAL CHAINS

In the previous section, we considered motion planning for a rigid object. In this section, we explore aspects of motion planning for a simple *articulated* object: a *robot arm*, which is simply an open polygonal chain. Aside from planning a specific path through an environment, we also discuss *reachability*, a motion-related question that is not path planning per se but which can be viewed from a configuration space perspective.

A typical robot arm is shown in Figure 7.8. This arm has a fixed base (the *shoulder*), to which are attached three rigid links, connected by motorized joints. This is known as a 6-DOF arm, because it has 6 *degrees of freedom*—independently controllable parameters that determine its movement. We will ignore all the (interesting) physical details and model a robot arm as a chain $C = (v_0, v_1, \ldots, v_n)$ of n *links*, each a rigid straight segment, connected at vertices v_i, each of which is a *universal joint* permitting a full range of motion. So, for us, a robot arm is an open polygonal chain. The shoulder v_0 we consider is pinned to the origin and the vertex v_n represents the *hand*.

Motion planning for an articulated object is conceptually similar to motion planning for a rigid object. Again the theme is to reduce the object to a point moving in a perhaps high-dimensional configuration space. The general idea is illustrated in the example shown in Figure 7.9(a). Here we have a 2-link arm in the plane, with its shoulder v_0 *pinned* (fixed in the plane). When dealing with *rigid* objects, our main concern dealt with collision avoidance of the object with obstacles. For articulated objects, we can also have a collision between the two links of the same object, where the articulated object is an obstacle to its own motion! However,

Figure 7.8. A robot arm.

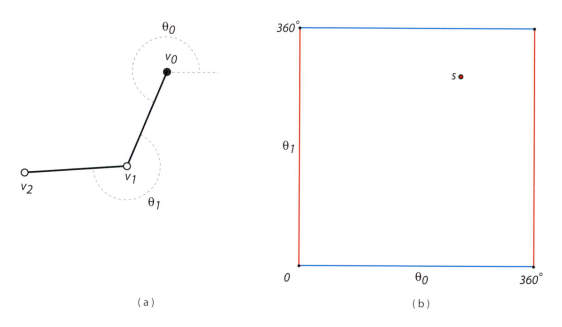

(a)

(b)

Figure 7.9. (a) A 2-link arm and (b) its configuration space.

for our initial model, we permit the links of a robot arm to pass freely through each other. We choose this to simplify the problem, but we should note that 2D arms can achieve this by staggering the links in separate layers parallel to a base plane.

We can represent the configuration of the arm in several ways, for example, by the coordinates of v_1 and v_2 in $\mathbb{R}^2$. But those four numbers are clearly not independent. A better representation of the configuration space is to consider *angles* rather than coordinates. It is clear that two angles suffice to determine the position of the arm, say θ_0 for the angle of v_0v_1 with respect to the x-axis, and θ_1 for the angle $v_2v_1v_0$ (as labeled in the figure). Then the point (θ_0, θ_1) determines the arm in a 2D configuration space $\mathcal{C}$. Figure 7.9(b) shows this configuration space as a square, where the horizontal axis is θ_0 and the vertical axis is θ_1, both ranging from $0°$ to $360°$. The particular configuration of the 2-link arm given in part (a) is associated to the marked point $s = (250°, 290°)$ in the space.

But this square does not tell us the whole story about $\mathcal{C}$. We know that a full $360°$ rotation of a link in the arm returns us back to the initial position. So the right and left edges (shaded red) of the square in Figure 7.9(b) (where $\theta_0 = 0°$ and $\theta_0 = 360°$) both represent the same configuration, and thus must be identified. Similarly, the top and bottom edges (shaded blue) of the square (where $\theta_1 = 0°$ and $\theta_1 = 360°$) must be identified. The resulting object is the actual configuration space of a 2-link arm. But what does this space look like after identifications of the edges? Figure 7.10 topologically shows that the actual space $\mathcal{C}$ is a torus, the surface of a donut! So each point on the torus corresponds to a particular configuration of a 2-link arm in the plane.

Let's consider now the case of the 2-link arm with obstacles. Figure 7.11(a) shows the arm above, again with its shoulder v_0 pinned, but now in an environment containing three polygonal obstacles. The configuration space is still based on a torus (corresponding to the two angles), but since there are obstacles on the plane, not all pairs (θ_0, θ_1) are present in the configuration space. Indeed, we must remove the points that cause the arm to intersect an obstacle. Figure 7.11(b) shows the

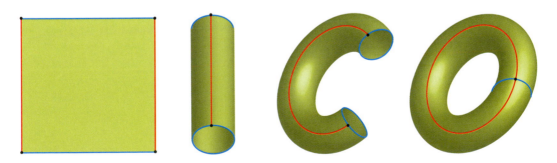

Figure 7.10. A square with identifications of its edges is a torus.

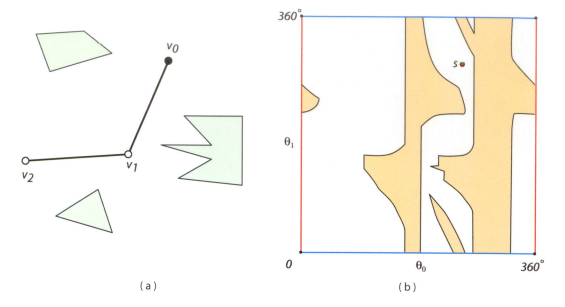

(a)   **(b)**

Figure 7.11. (a) A 2-link arm with obstacles and (b) its configuration space.

configuration space for the 2-link arm, where the shaded regions must be removed from the torus. So now we have reduced planning the motion of the arm to finding a path between initial and final points in this toroidal configuration space. The general motion-planning algorithms previously described apply to this situation as well.

What happens when we assume the two links of the arm cannot pass freely through one another, where the articulated object is an obstacle to its own motion? For a 2-link arm, the restriction is simply that the second link cannot cross through the first link, which excludes the $\theta_1 = 0° = 360°$ in our parametrization. For an n-link arm, there are $\binom{n}{2}$ possible link-link collisions to be avoided, and applying the general roadmap algorithm leads to a complexity of $2^{O(n^2)}$. Although this is formidable, real robot arms have only a small number of degrees of freedom, usually $n \leq 10$. And path planning in the resulting n-dimensional configuration spaces is in fact used in implementations.

However, there are robot arm-like structures for which n is significantly larger. One example is snake or serpentine robots, used for search and rescue missions or for surgery, which might have $n = 30$. A rather different and quite important example is the backbone of a protein molecule, where values of $n = 10,000$ are reached. Exponential algorithms are useless in these cases, and approximation techniques have been developed, one of which we discuss later.

Exercise 7.7. *How many connected components does the configuration space of Figure 7.11(b) have? How about if we do not allow the two links to pass freely through one another?*

In our discussion below, we consider robot arms in the plane without obstacles, where we again permit the links of a robot arm to pass freely through each other. Among the simplest questions one can ask about such arms concerns *reachability*: Given a point p in the plane and an arm A specified by its link lengths $[\ell_1, \ldots, \ell_n]$, with its shoulder joint v_0 pinned at the origin, can A reach p? In other words, can arm A be configured so that $v_n = p$? Note that this is not a motion-planning question in two senses. First, we are asking for a YES/NO answer rather than for a *path*, and second, the final configuration is only weakly specified in that we only care about the location of the hand v_n and not the other joints of A.

It may seem the answer is obvious, where p is reachable if and only if the distance from p to the origin is not greater than the sum $\ell_1 + \ell_2 + \cdots + \ell_n$. But this is not in general true: it may be that points near the shoulder are inaccessible due to the particular lengths of the links in the arm. Let the *reachability region* for an arm be the set of all points that v_n can reach.

Theorem 7.8. *The reachability region R for an arm A of link lengths $[\ell_1, \ldots, \ell_n]$ is an annulus[3] centered at v_0, the region between two concentric circles in $\mathbb{R}^2$. The outer radius of the annulus is*

$$r_o = \ell_1 + \ell_2 + \cdots + \ell_n,$$

and the inner radius is

$$r_i = \begin{cases} L - M & \text{if} \quad L > M, \\ 0 & \text{if} \quad L \leq M, \end{cases}$$

where L is the length of the longest link in the arm and M the sum of the lengths of all the other links.

Proof. We first prove that R is an annulus and then compute the radii. The annulus proof is by induction. An arm of $n = 1$ link can reach the points on a circle of radius ℓ_1 centered on v_0, which is an annulus by our definition. Suppose now that the lemma holds for all arms of up to $n - 1$ links. Let $A' = [\ell_1, \ell_2, \ldots, \ell_{n-1}]$ be the arm A with the last link removed. By the induction hypothesis, the A' reachability region R' is an annulus centered on v_0. Let $S(r)$ be the circle of radius r centered at the origin. Then the reachability region for A is the Minkowski sum $R' \oplus S(\ell_n)$, the union of circles of radius ℓ_n centered on every point of R'. This is again an annulus: with outer radius larger by ℓ_n, and inner radius reduced either by ℓ_n or to zero (if the origin can be reached by v_n). See Figure 7.12 for a diagram of (a) the annulus R' and (b) the newly formed annulus R.

---

[3] We include under the term "annulus" two degenerate situations: when the annulus is a circle ($r_i = r_o$) and when the annulus is a disk ($r_i = 0$).

(a)

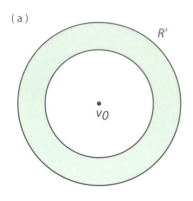

(b)

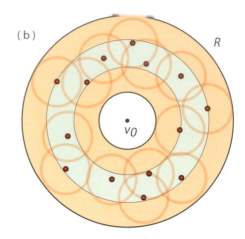

Figure 7.12. (a) The annulus R' centered at v_0 and (b) the annulus R formed by the Minkowski sum of R' and the circle of radius ℓ_n.

We now turn to computing the radii. The outer radius is easy: the furthest reach of the arm is achieved by straightening each joint, stretching the arm out straight. So r_o is the sum of the link lengths in A as claimed. To compute r_i, first note that the region of reachability R of an arm A is independent of the order of the links in the arm. This is because v_n can be reached by summing successive vectors,

$$v_n = v_0 + (v_1 - v_0) + (v_2 - v_3) + \cdots + (v_n - v_{n-1}),$$

and vector addition is commutative. So shuffling the links leaves R unchanged.

Let $L = \ell_k$ be a longest link in arm A. Reorder the links of A by moving link k to the front, incident to the shoulder. This does not alter the reachability region, but helps intuition. Let M be the sum of the lengths beyond this new first link. The reshuffling makes it clear that if $M < L$, then the hand cannot reach the shoulder. The closest v_n can get to v_0 is $L - M$, and this is r_i. Figure 7.13 shows this situation where the longest link (in red) is of length L, and the lengths of the other links (in blue) sum up to M. When $M \geq L$, then v_n can reach v_0, and then the annulus becomes a disk, and $r_i = 0$. $\qquad\square$

So the decision question we posed at the beginning of this section is answered easily, in $O(n)$ time: compute r_i and r_o and check if $r_i \leq |p| \leq r_o$. Note that knowing the answer is YES does not immediately tell us how to configure the arm to reach p. It turns out this can also be achieved in $O(n)$ time.

The above argument works not just for linkages in the plane, but for linkages in any dimension: The natural generalization of an annulus in the plane — the region trapped between two circles — is extended to the region trapped between two higher-dimensional spheres. One might

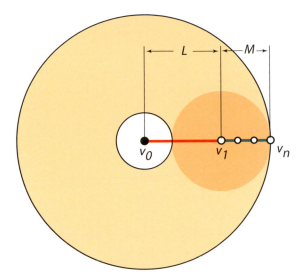

Figure 7.13. When $L > M$, the inner radius is $r_i = L - M$.

be tempted to conclude that the reason these reachability questions are so easy in comparison to motion planning is that we excluded obstacles. The next section shows that even some configuration questions for arms in the absence of obstacles are very difficult.

Exercise 7.9. *The last sentence of the proof of Theorem 7.8 claims that if $M \geq L$, then v_n can reach v_0. Prove this intuitively plausible claim.*

Exercise 7.10. *Design an algorithm to find at least one configuration for an n-link arm to reach a given point p within its reachability region. Try for $O(n)$ time complexity.*

Exercise 7.11. *Argue that if an arm can reach a point p, it can reach p with at most two joints* kinked, *where a joint is kinked if its joint angle is different from π, i.e., it is not straightened.*

Exercise 7.12. *Describe the shape of the reachability region of a 2-link arm with joint angle constraints $\theta_0 \in [\alpha_0, \beta_0]$ and $\theta_1 \in [\alpha_1, \beta_1]$. Here θ_0 and θ_1 are defined as in Figure 7.9(a).*

7.3 RULERS AND LOCKED CHAINS

Suppose one wants to stow an *n*-link robot arm in a small space, for example, in the Space Shuttle storage bay. The natural method is to fold it flat, perhaps alternating the joint angles between fully turned clockwise and fully turned counterclockwise. However, this only produces a compact configuration if the links are all about the same length. For an arbitrary n-link arm, it is less clear how to fold it flat compactly.

An alternative formulation of the problem is obtained by viewing the arm as a strange *carpenter's ruler*, with measuring segments of differing lengths. One wants to fold the ruler flat so that, end-to-end, it has the smallest total length possible for its link lengths. Note that this goal is neither a motion-planning question nor a reachability question. More formally, we have the following:

Ruler Folding Problem. *Given an integer L and a polygonal chain with links of integer lengths $[\ell_1, \ldots, \ell_n]$, can the chain be folded flat— reconfigured so that each joint angle is either 0 or π—so that its total folded length is less than or equal to L?*

In 1985, John Hopcroft, Deborah Joseph, and Sue Whitesides established that this problem is NP-complete, an intractable algorithmic problem, as explained in the Appendix. Although we have mentioned NP-completeness several times, we have yet to show how one establishes a particular problem to be in this intractable class. The usual method of proving NP-completeness is to reduce our problem to a known intractable problem. Here we choose the SET PARTITION problem as the target:

Set Partition Problem. *Given a set of n positive integers $S = \{x_1, \ldots, x_n\}$, does there exist a partition of S into sets $A \subseteq S$ and $B \subseteq S$ so that*

$$\sum_{x_i \in A} x_i = \sum_{x_j \in B} x_j.$$

This problem has been proved NP-complete and so no polynomial-time algorithm is known for it. Now the task is to show that if one could solve the RULER FOLDING problem quickly (i.e., in polynomial time), then one could solve any instance of SET PARTITION quickly. And that contradiction establishes that the ruler-folding problem is itself NP-complete.

Theorem 7.13. *The RULER FOLDING problem is NP-complete.*

Proof. We start from an arbitrary instance of SET PARTITION. Let $S = \{x_1, \ldots, x_n\}$ be a set of positive integers and let s be the sum of all the elements in S. Now we construct an instance of the ruler-folding problem that will "solve" the SET PARTITION instance. Construct a ruler R consisting of links of lengths

$$[2s, s, x_1, \ldots, x_n, s, 2s]$$

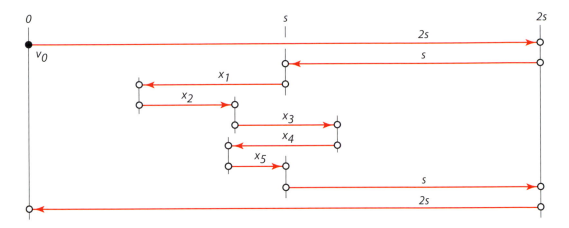

Figure 7.14. A ruler folding reduction, where $S = \{23, 15, 16, 17, 9\}$ and $x_1 + x_4 = x_2 + x_3 + x_5$.

connected at vertices $(v_0, v_1, \ldots, v_{n+4})$. The claim is that R can be folded into a length of at most $2s$ if and only if the instance of SET PARTITION has the answer YES.

Consider the links directed from v_i to v_{i+1}, and fold the ruler along the real line, with $v_0 = 0$ and $v_1 = 2s$. The two vertices v_2 and v_{n+2} must both lie at s if all is to fit within the interval $[0, 2s]$. This forces the links between to consume zero total displacement. So the sum of the leftward-pointing links must equal the sum of the rightward-pointing links. Figure 7.14 shows an example with 9 links; the endpoints at each of the rods have been separated for clarity.

Therefore R folds into $2s$ if and only if $\{x_1, \ldots, x_n\}$ can be partitioned into two equal-length halves; in the figure, we have $x_1 + x_4 = x_2 + x_3 + x_5$. It is clear that it cannot fold to less than $2s$ because there are links of that length. So we have established the claim for $L = 2s$. $\square$

The import of this theorem is that no one knows a method for solving the ruler-folding problem better than trying all 2^n foldings (there are two choices of folding direction at each of n joints) and seeing whether any result in a length less than L. For $n = 100$, trying $2^{100} \approx 10^{30}$ foldings is infeasible even on the fastest computer.

The RULER FOLDING problem asks for finding a point in the configuration space with particular properties, but the configuration-space viewpoint does not play a significant role in this problem. Now we turn to a question for which it does play a central role. Recall that whether a path exists between configurations s and t depends on whether s and t lie in the same connected component of the configuration space. A fundamental question asks whether all of the configuration space $\mathcal{C}$ is connected, that is, is it a single connected component or does it have several disconnected components? If $\mathcal{C}$ is connected, then any configuration can move to any other configuration, whereas if $\mathcal{C}$ is disconnected, then some configurations are inaccessible from some others.

Thus far we have been assuming that the links of our robot arm can pass freely through each other. In the subsequent discussion, let $C = [\ell_1, \dots, \ell_n]$ be a polygonal chain that is *not* permitted to self-intersect in an otherwise obstacle-free environment. If the configuration space $\mathcal{C}$ is disconnected, we say that the chain can *lock*, because it can get stuck in a configuration component from which it cannot escape. Do there exist locked chains? This natural question was posed in the 1970s and independently re-posed several times in succeeding decades until it was finally resolved in the late 1990s. The answer actually depends on the ambient dimension in which the chains exist. For chains in the plane $\mathbb{R}^2$, there are no configurations that lock. In other words, the configuration space of motions of the chain is connected. (Indeed, this turns out to be true in $\mathbb{R}^d$ for all $d \geq 4$.) There are locked chains in 3D, however. We will prove the $\mathbb{R}^3$ result, but can only hint at the more difficult $\mathbb{R}^2$ theorem.

Let's start with three dimensions. It is of course easy to lock a *closed* chain in 3D by tying it into a knot. But it is less immediate for open chains, our concern here. There is a relatively easy example of a 5-link locked open chain, dubbed the *knitting needles*, shown in Figure 7.15. Even though there is a sense in which it is obvious that this chain is locked, finding a clean proof has proved more delicate. Several proofs appeared after 1998 and we present one of them now, which invokes basic knot theory as its final step.

Theorem 7.14. *The knitting needles chain $K = (v_0, \dots, v_5)$ in Figure 7.15 cannot be made straight when the end links are longer than the lengths of the middle links combined.*

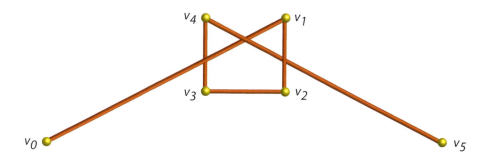

Figure 7.15. The *knitting needles* locked chain.

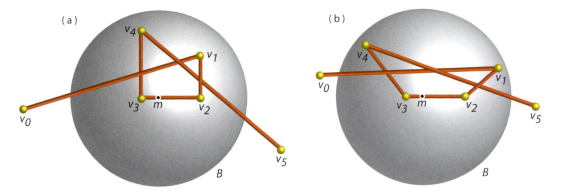

Figure 7.16. (a) Ball B is centered on the midpoint m along (v_1, v_2, v_3, v_4). (b) A reconfiguration of the chain illustrates that v_1 and v_4 remain interior to B (by at least ε) and v_0 and v_5 stay exterior (by at least 2ε).

Proof. The proof starts at the configuration shown in the figure and argues that the chain cannot be made straight. This then establishes that $\mathcal{C}$ has at least two disconnected components. Let $\ell_k = |v_{k-1}v_k|$ be the length of the kth link, and let $L = \ell_2 + \ell_3 + \ell_4$ be the total length of the short central links. We set ℓ_1 and ℓ_5 below. Let $r = L/2 + \varepsilon$, for small $\varepsilon > 0$, and center a ball B of radius r on the midpoint m of the three central links, that is, m is $L/2$ from both v_1 and from v_4 along the chain. Figure 7.16(a) provides a picture. By construction, we have $\{v_1, v_2, v_3, v_4\} \subset B$ during any reconfiguration of the chain. Now choose ℓ_1 and ℓ_5 to be at least $2r + \varepsilon = L + 3\varepsilon$. Because this length is greater than the diameter of B, the vertices v_0 and v_5 are both necessarily exterior to B during any reconfiguration, such as the one in Figure 7.16(b).

Assume now that the chain K can be straightened by some motion. We will reach a contradiction. Because of the separation maintained between $\{v_0, v_5\}$ and $\{v_1, v_2, v_3, v_4\}$ by the boundary of B, we could have attached a sufficiently long unknotted string s from v_0 to v_5 exterior to B that would not have hindered the unfolding of P. But this would imply that the trefoil knot $K \cup s$ can be straightened (without self-crossings) into the trivial knot. We have reached a contradiction and therefore, K cannot be straightened. ☐

If all three central links ℓ_2, ℓ_3, ℓ_4 have the same unit length, then the long first and last "needle" links must have length more than 3. This length ratio is critical in that the configuration space $\mathcal{C}$ has just a single component if the needles are not long enough. It is also known that all chains of fewer than five links are unlocked, regardless of link lengths. However, this basic question remains open:

UNSOLVED PROBLEM 27   3D Unit Chains

Can a chain in $\mathbb{R}^3$ lock if its link lengths are all the same?

We mentioned that motion planning for an n-link arm can be accomplished by a general algorithm that runs in $2^{O(n^2)}$ time. A modification of this algorithm permits counting the components and so deciding whether any given n-link chain in $\mathbb{R}^3$ is locked.

UNSOLVED PROBLEM 28   3D Locked Chains

Find a polynomial-time algorithm to decide whether or not a 3D chain is locked.

Exercise 7.15. *Show that any chain of four links is unlocked.*

Exercise 7.16. *Argue that any simple open chain lying in a plane is not locked in 3D, in the sense that it can be reconfigured to a straight configuration of one link after another lying on a line in $\mathbb{R}^3$.*

★ **Exercise 7.17.** *Show that any collection of chains in 3D, each of which consists of just two links, can be separated arbitrarily far from one another (without any link crossing another).*

Let's now consider the situation of chains in the plane, which is a far more difficult question. After a long pursuit, it was finally established in 2003 by Robert Connelly, Erik Demaine, and Günter Rote that chains cannot lock in 2D. More precisely, the result is as follows:

Theorem 7.18 (Carpenter's Rule). *Every collection of chains in the plane has a motion to a configuration in which every outermost open chain is straightened and every outermost closed chain is convex.*

This result includes as a special case that every single open chain C can be straightened. Let s and t be any two simple configurations of C, and let c be the straightened configuration of C. Because we can move from s to c and from t to c, we can move from s to t via c by reversing the motion from t to c. The straightened configuration c is called a *canonical configuration*, a standard form that can serve as a universal stopping point between any two configurations. In a similar way, Theorem 3.22

(a)

(b)

(c)

(d)

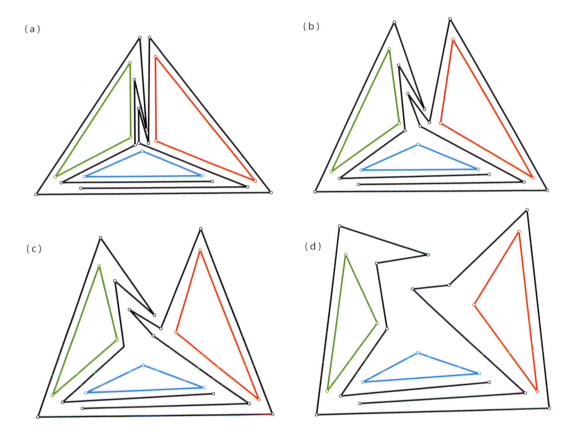

Figure 7.17. Four frames in the unlocking of one open chain and three closed (polygonal) chains. Figure courtesy of Erik Demaine.

showed the flip graph — our *discrete* configuration space of triangulations — is connected by proving that any triangulation can be made into a "canonical triangulation" (which, in this case, came from the incremental algorithm).

Notice that the theorem includes convexifying closed chains, that is, polygons. The reason for the "outermost" qualifications in the statement is that it is not always possible to straighten or convexify a chain if it is curled up inside a small enclosing polygon. Figure 7.17 illustrates four frames toward the unlocking of three closed chains (colored) entangled with one open chain. This example is one that superficially might seem to be locked, but clearly it is not, as shown by part (d). Indeed many researchers proposed potentially locked chains (or collections of chains) before the issue was finally resolved.

A full proof of Theorem 7.18 is quite complicated, so we content ourselves with mentioning a few aspects. One key insight is that there always exists an *expansive motion*, one in which the distance between every pair of vertices either increases or stays the same. An expansive motion guarantees simplicity throughout because two segments can only cross by one or more vertices decreasing their distance. A differential equation for expansive vector motions for each vertex can be formulated and tracking its solution leads to an unlocking motion. A numerical solution of the differential equation has been implemented, which was used to generate the motion in Figure 7.17. Since the original proof, two more unlocking algorithms have been developed, one using pseudotri-angulations (Section 3.5) and the other minimizing an energy function. The latter has led to a novel graphics morphing algorithm, yet another instance of a practical application affected by the pursuit of a purely theoretical question.

Exercise 7.19. *Prove that if two non-adjacent links of a simple open chain move to intersect each other, then some pair of vertices of the chain must strictly decrease their separation.*

We close this section by exploring an algorithm developed for protein folding. The protein folding problem is one of the most important unsolved problems in all of science. A protein molecule is composed of a chain of amino acids residues joined by peptide bonds. Its central *backbone* consists of three atom vertices per residue, with adjacent atoms connected by bond links. We can crudely model the backbone by a polygonal chain. A typical protein has between $n = 100$ and $n = 1000$ atoms along its backbone, but some such as the muscle protein titin have $n = 30,000$ atoms.

Natural proteins have two remarkable folding properties. First, they appear to have a unique minimum-energy folded state, determined solely by their "primary structure," that is, the sequence of specific amino acids along the backbone. Second, they curl up unerringly to this state in about one second, as illustrated in Figure 7.18. The *protein folding problem* is to predict the unique 3D folded configuration (the "tertiary structure" in biochemical terminology) from the primary structure. Because the functionality of a protein is largely determined by its shape, the ability to predict the folding from the sequence of acids would enable quick, targeted drug design. If we ignore the biochemistry and biophysics of protein folding, then protein folding can be viewed as reconfiguring a polygonal chain, perhaps with the "side chains" attached to the backbone. In this view, it is a motion planning problem, but one with n so large that the general exponential motion planning algorithms are useless.

The unfeasibility of using the roadmap algorithm (recall from Section 7.1 that this is the fastest known) for objects with a large number of degrees of freedom led, in the early 1990s, to two groups independently inventing a new structure called the *probabilistic roadmap* (PRM), which was immediately successful in increasing the range of accessible degrees-of-freedom values. This has developed into an approach called

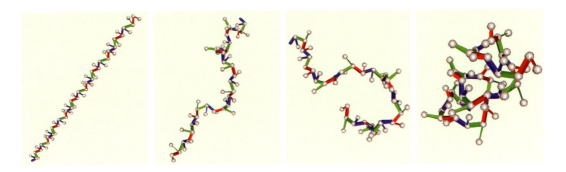

Figure 7.18. Folding snapshots of a protein molecules consisting of 10 amino acids. Here the side chains are attached to the backbone. Figure courtesy of Nancy Amato and Guang Song.

sampling-based motion planning, which circumvents exact construction of the configuration space. Here we sketch the original PRM technique.

Let $\mathcal{C}$ be the configuration space of a robot arm (or other movable object) with k degrees of freedom. The PRM motion planning algorithm consists of two phases: the roadmap construction and the query phase. The roadmap is constructed by generating random points and checking to see if they are in the configuration space $\mathcal{C}$, and if so, connecting nearby points into a graph structure G with a "local planner." The query phase takes start and goal configurations s and t, connects each to G, and uses G to construct a path from s to t. The folding shown in Figure 7.18 was computed using this algorithm. One of the challenges of the PRM technique is generating sufficiently many random points to accurately capture the connectivity of $\mathcal{C}$, which may have narrow, convoluted tunnels. Usually some heuristic (but theoretically guided) method is used to increase the density of sample points in these difficult regions. This is an active area of current research.

7.4 POLYGON SPACES

The previous section focused on various aspects of configuration spaces for open polygonal chains. Now we turn to closed polygonal chains, specifically, polygons in 2D. All edges are considered rigid links whose lengths are fixed, and all vertices are universal joints. We ignore intersections, as in our study of robot arm reachability and in contrast to the investigation of locked chains. Finally, we focus on the topological structure of the configuration space, in contrast to the various other properties we have examined previously. As a word of caution, the material in these remaining few sections of the book are advanced, forcing us to provide sketches and general overviews of many ideas. But even though these topics require a certain amount of mathematical sophistication, we believe it is worth the effort to gain glimpses of these worlds.

We first review the notion of "topological structure," first touched on in Section 5.6 and Section 6.2. Two spaces are topologically equivalent (under homeomorphism) if one can be distorted to the other without tearing or gluing. So the sphere and the torus are topologically distinct; the symbols S^2 and T^2 are used to describe the topological type of a sphere and a torus, respectively. Both of these spaces are *surfaces*, which means that they look like $\mathbb{R}^2$ in the neighborhood of every point. The exponent on S^2 and T^2 reflects the intrinsic dimension of the manifold rather than the dimension of the space in which the manifold might be embedded. Thus S^1 is the topological type of a circle, since locally it looks like a 1D line.

Let's review the topology of some of the configuration spaces so far considered. The space for a polygon translating in 2D is $\mathbb{R}^2$, and that for a polyhedron translating in 3D is $\mathbb{R}^3$. A polygon translating and rotating in $\mathbb{R}^2$ leads to the manifold $\mathbb{R}^2 \times S^1$, the Cartesian product of the translation-only space $\mathbb{R}^2$ with S^1 for the one rotation angle. Each point in this space keeps tracks of three pieces of data: two entries for the amount of translation (a plane's worth of choices) and one entry for the amount of rotation (a circle's worth of choices). Thus $\mathbb{R}^2 \times S^1$ is a 3-manifold. A polyhedron translating and rotating in 3D has six degrees of freedom, and its configuration space is $\mathbb{R}^3 \times SO(3)$, where $SO(3)$ is the "special orthogonal group" that captures the three degrees of rotational freedom.

The space for a 1-link planar robot arm is the circle S^1. For a planar 2-link arm, the space is the torus T^2, as previously illustrated in Figure 7.9(b) and Figure 7.10. From these images, we see that T^2 can be identified as the product $S^1 \times S^1$. By a similar argument, one can show that the space for an n-link planar arm is the n-dimensional torus T^n, the product of n circles. Our (limited) goal in this section is to show that the topology of the configuration space of an equilateral pentagon in the plane is a genus-4 surface.

Exercise 7.20. *Show that the configuration space for a 3-link arm in the plane is T^3.*

Exercise 7.21. *What is the topology of the configuration space of a 2-link arm if the point v_0 is not pinned to the plane?*

Exercise 7.22. *What is the topology of the configuration space of a triangle that is nowhere pinned to the plane?*

Our *polygonal system* $\mathcal{P}$ is given by n link lengths $[\ell_1, \ldots, \ell_n]$. A *realization* P of $\mathcal{P}$ is specified by actual coordinates of the polygon vertices $v_0, \dots, v_{n-1}$. (We continue to use the term "polygon" to include nonsimple closed polygonal chains.) Stringing together these coordinates makes a point in $\mathbb{R}^{2n}$, since each vertex requires two pieces of data. We seek to understand the topology of the subset of $\mathbb{R}^{2n}$ consisting of all realizations P of $\mathcal{P}$, that is, the configuration space $\mathcal{C}$. One step in this direction is to find the dimension of $\mathcal{C}$ as a function of n. In other words, how many degrees of freedom of motion does an polygon with n vertices have?

In Section 7.2, we considered *open* chains. For all such chains, we always pinned the shoulder v_0 onto the plane, fixing its motion. The other joints were allowed to flex and turn. For *closed* chains, in the case of polygons, we fix not just a vertex v_0 but an entire edge ℓ_1 onto the plane. Thus the first vertex v_0 and second vertex v_1 of the polygon (with the edge ℓ_1 between them) are pinned; in particular, we can place vertex v_0 at the origin $(0, 0)$ and vertex v_1 at $(\ell_1, 0)$ on the x-axis.

The simplest examples of polygonal linkages are triangles. Our polygonal system $\mathcal{P}$ is then given by three link lengths $[\ell_1, \ell_2, \ell_3]$. If the first edge is pinned, fixing two vertices, the third vertex is immediately determined up to reflection (because a triangle is rigid). Thus the configuration space for a triangle is simply two points, corresponding to v_2 above or below $v_0 v_1$. In this case, the dimension of $\mathcal{C}$ is zero, even though $\mathcal{C}$ is a subset of $\mathbb{R}^2$ (if the base is fixed).

What about quadrilaterals? Consider the planar quadrilateral depicted in Figure 7.19(a). Again fix v_0 at the origin and v_1 at $(\ell_1, 0)$ on the x-axis. Now the motions of v_2 are restricted to rotations about a circle of radius ℓ_2 centered on v_1. Similarly, v_3 is forced to rotate about a circle of radius ℓ_4 centered at v_0. There is one more restriction needed. The link between v_2 and v_3 forces v_3 to be on a circle of radius ℓ_3 centered at v_2. Figure 7.19 depicts two configurations of this quadrilateral linkage, where v_2 and v_3 lie on the boundaries of the blue and red disks respectively. The dashed circle is of radius ℓ_3 centered at v_2.

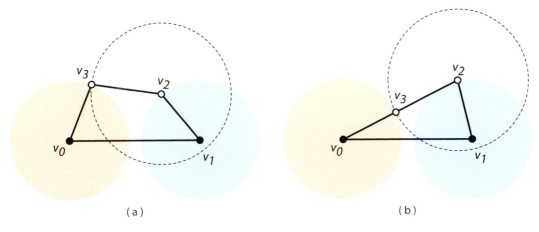

(a)   (b)

Figure 7.19. Reconfiguration of a quadrilateral.

Therefore, v_3 must lie at the intersection of the two circles centered on v_0 and on v_2. Because two circles (generically) intersect at two points, each position of v_2 (generically) results in two possible positions for v_3. As v_2 moves through its range of motion, v_3 is determined by this intersection point, until the circles become tangent and the two circle intersection points become one, as in Figure 7.19(b). From here, v_3 can track the other intersection point. This example makes clear that $\mathcal{C}$ is one-dimensional, as only v_2 has freedom of motion—the position of v_3 varies continuously as a function of the position of v_2. Indeed, the topology of this configuration space $\mathcal{C}$ turns out to be just a circle! Figure 7.20 depicts this space (in red), where the configurations of the linkage corresponding to eight points of $\mathcal{C}$ are drawn. In particular, note the symmetry of antipodal points on the circle. As we travel around the circle $\mathcal{C}$, the motion of the linkage looks somewhat like locomotive wheels. For a quadrilateral system different from the linkage lengths given by Figure 7.19(a), the configuration space $\mathcal{C}$ might not necessarily be a circle.

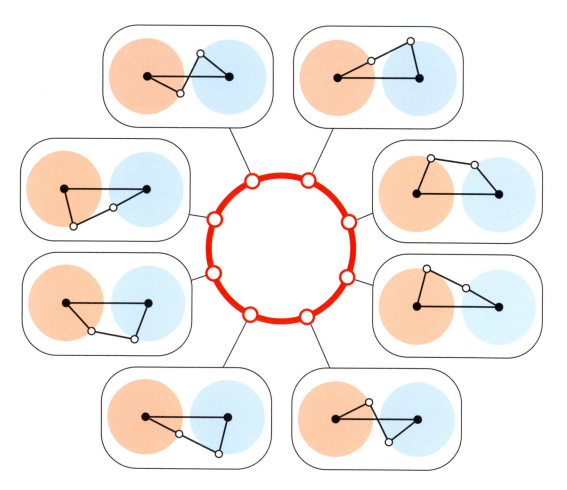

Figure 7.20. The configuration space of this quadrilateral polygon is a circle.

Exercise 7.23. *Let* $\mathcal{P} = [5, 4, 4, 1]$ *be the lengths of a quadrilateral system. Show that its configuration space consists of two disconnected circles.*

Exercise 7.24. *Argue that the configuration space for a parallelogram with unequal side lengths is a circle.*

Exercise 7.25. *Argue that the configuration space for a rhombus, that is, an equilateral quadrilateral* $\mathcal{P} = [1, 1, 1, 1]$, *consists of three circles, each pair of which intersect in a single point!*

The configuration space of a pentagon is much more complicated than that of a quadrilateral. Just by analogy with the configuration space of triangles (0D) and quadrilaterals (1D), you can guess the dimension here is 2. This is supported by a similar construction, where we fix v_0 and v_1 and rotate v_2 and v_4 on arcs of circles centered on v_1 and v_0, respectively. This determines v_3 just as for a quadrilateral. So we have two degrees of freedom: rotation of v_2 about v_1 and of v_4 about v_0. In general, the dimension of the configuration space $\mathcal{C}$ for a polygon with n vertices is $n - 3$. Because pentagon configuration spaces are 2D surfaces, they can be visualized, although we should remember they are surfaces embedded in $\mathbb{R}^6$—fixing the first edge leaves three vertices to vary. However, the intersecting-circles viewpoint used for the quadrilateral case yields little insight into the topological structure for the pentagonal linkages.

To proceed further, we will employ a powerful tool known as *Morse theory*, developed by the differential geometer Marston Morse in the 1930s. Morse theory is a large topic, and we will only hint at enough of it to sketch out how it can be used to determine the topology of the configuration space of pentagonal linkages. We start with a *Morse function* $f : S \to \mathbb{R}$ which maps every point on a surface S to a real number. It is convenient (and conventional) to imagine a 2D surface S embedded in $\mathbb{R}^3$, with $f(p)$ the z-coordinate of a point p of S. For example, S might be a torus, a donut stood on its end, as illustrated in Figure 7.21(a). Now we slice S by a plane Π_z at each z-value, imagining z continuously decreasing from above S to below S. The intersection $\Pi_z \cap S$ changes its topology only at certain *critical points*. In this example, the critical points occur when Π_z first touches the top of the torus, next when the intersection first becomes two components at the top of the hole, next when those components merge at the bottom of the hole, and finally when Π_z last touches the bottom of the torus. Part (b) shows certain slices by the plane Π_z.

The critical points can be identified as having degenerate first derivatives (partial derivatives with respect to vertex coordinate variables),

(a)   (b)

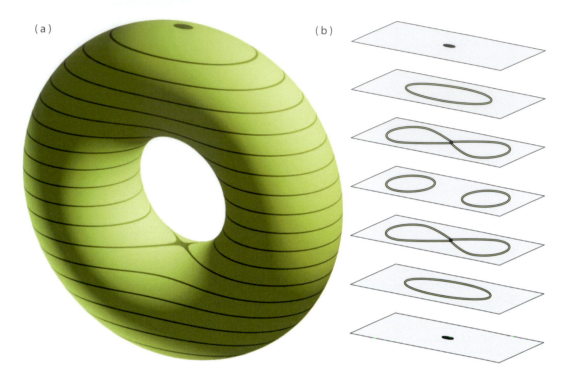

Figure 7.21. (a) Slicing a torus at different heights. (b) Certain slices by the plane Π_z, where every other height depicted here is a critical point.

and each critical point can be assigned an *index* based on the second derivatives there. It is this index that determines the type of topological change occurring at the critical point. We will explain none of this but only claim that, for the torus example, index 2 indicates that a circle appears, index 1 indicates that a split or merge of components occurs, and index 0 marks the disappearance of a circle. And Π_z passes through four critical points (of index 2, 1, 1, and 0) as it sweeps over the torus.

With this background, let's apply Morse theory to the case of pentagonal linkages. Let $\mathcal{P}$ be the polygonal system $[\ell_1, \ldots, \ell_5]$. The first thing we will need is a Morse function. This can be accomplished by pinning v_0 to the origin and v_1 to $(z, 0)$ along the x-axis, but leaving the first link length z variable. The Morse function $f : \mathcal{P} \to \mathbb{R}$ is simply $f(P) = z$. For a large value of z, the configuration space $\mathcal{C}$ will be empty: the sum of lengths $\ell_2 + \cdots + \ell_5$ will be less than z, not allowing the polygon to close up. As z decreases in value, we will track the topological changes along the way using Morse theory. By the time $z = \ell_1$, we will know the topological structure of $\mathcal{C}$.

It turns out that the critical points of f all occur when the links all lie on the x-axis. We will not justify this claim but it does make intuitive sense that the derivatives are degenerate precisely here. In contrast to the simple situations with quadrilaterals, the topological structure of $\mathcal{C}$ for a pentagon varies dramatically depending on the lengths $[\ell_1, \ldots, \ell_5]$. To

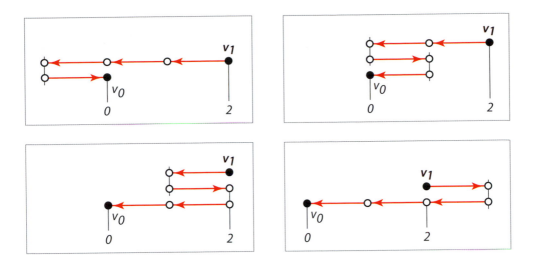

Figure 7.22. The four critical configurations of a unit pentagon at $z = 2$. Three links are pointing leftward, and one rightward.

avoid this complication, we only examine the equilateral pentagon, with $\ell_i = 1$ for all five links, calling it the *unit pentagon*.

This leaves us with five critical points, the five distinct ways to lay down the first four links of P on the positive x-axis: One critical point is when all four links stretched out straight, reaching $z = 4$. Four other critical points appear when three links are pointing leftward and one rightward, reaching $z = 2$; Figure 7.22 shows these configurations. Note that there are four possibilities here, depending on which link points rightward. The only other possible way in which the linkage folds flat is when the first four links reach $z = 0$; in this case, the last link must be $\ell_5 = 0$, which contradicts our assumption that $\ell_i = 1$.

Although we know the configuration space of a pentagonal linkage is a surface, we obtain a *3-manifold* when ℓ_5 is considered as a variable z, having one more degree of freedom. The indices of the critical points of a manifold have a different topological interpretation than in our 2D torus example above. For example, the index-2 event when the 2D plane Π_z first intersects the surface in a circle now corresponds to an index-3 event when the 3D hyperplane Π_z first intersects the 3-manifold in a sphere. In particular, we have this index-event correspondence, given in the table below, which we ask the reader to accept on faith.

Index Number	Event
3	Sphere appears
2	Attach a handle
1	Detach a handle
0	Sphere disappears

As explained above, we start at a large value for z. For our unit pentagon, recall that for all $z > 4$, the configuration space is empty. At $z = 4$, the surface becomes a sphere, an index-3 event. (This corresponds to the case when all four links are stretched out straight, reaching $z = 4$.) No further topological changes occur until $z = 2$, when we hit four critical values, each of index 2; these values are the ones shown in Figure 7.22. According to the table above, each of these causes a *handle* to be attached to the sphere, a loop similar to a coffee cup handle. A sphere with one handle attached is topologically a torus, and a sphere with four attached handles is a genus-four surface. Finally, z moves to $z = 1 = \ell_5$ without further topological changes.[4] So the configuration space $\mathcal{C}$ for a unit pentagon linkage is a genus-four surface, homeomorphic to Figure 6.6, as originally claimed.

One can get some sense of "walking around a handle" in the configuration space through a sequence of reconfigurations similar to what is shown in Figure 7.20. Indeed, if the pentagon is not equilateral, its configuration space might be a genus-two or genus-three surface, or even two disconnected tori. And these are just the nondegenerate situations. If spaces with *singularities* are included (such as those with "pinchpoints" resulting in nonmanifolds), there are 19 distinct topological types of pentagon configuration spaces!

Exercise 7.26. *Find lengths for a pentagonal system whose configuration space is disconnected.*

★ **Exercise 7.27.** *Prove that the configuration space of the unit pentagon has genus 4 by constructing a combinatorial "cell decomposition" of the space and applying Euler's formula.*

★ **Exercise 7.28.** *Call a pentagon degenerate if it can be flattened to lie in a line. Select a particular degenerate pentagon, and argue that its configuration space has a singularity, a point at which it fails to be a manifold.*

★ **Exercise 7.29.** *Find a condition on the lengths of the edges of a polygon of n vertices that implies that the configuration space is disconnected. Start by generalizing Exercises 7.23 and 7.26 for a condition for quadrilaterals and pentagons.*

---

[4] There are no critical points of index 1 for unit pentagons before $z = \ell_5 = 0$.

7.5 PARTICLE COLLISIONS

Throughout this chapter we have looked at configuration spaces of rigid objects, of linkages, and of polygons. We close by looking at one of the most important kinds of configuration spaces in mathematics, the space of particles. Here our robot equivalents are simply *points* in space, having no width or height. In what ways can these point robots (henceforth, *particles*) interact with one another? As with the other topics in this chapter, there is a huge field of study related to this topic. We will be content to focus our attention on the configuration space of particles on the unit interval.

First some notation. Let $I = [0, 1]$ be the unit interval in $\mathbb{R}$ with *fixed* particles at the two endpoints. The configuration space of n particles on this interval is denoted as $C_n(I)$. These n particles are allowed to move along this interval and are allowed to touch particles adjacent to them. Since each particle's position corresponds to a value on the real number line, the configuration space for n particles is the set of points $(x_1, x_2, \ldots, x_n)$ in $\mathbb{R}^n$ that satisfy the inequality

$$0 \leq x_1 \leq x_2 \leq \cdots x_n \leq 1. \tag{7.1}$$

But this equation does not help us understand the structure of this configuration space.

In order to visualize this, let's start with just one particle x, the case $C_1(I)$. Since the position of x can be anywhere inside I, the configuration space is the interval itself, as shown in Figure 7.23. This figure also

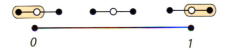

Figure 7.23. The configuration space $C_1(I)$ along with a labeling of the chambers.

provides a labeling of the three "chambers" of this space: the interior $(0 < x < 1)$ and the two vertices $(x = 0$ and $x = 1)$ where the particle has touched the fixed end-particles of I. Notice the use of a "tube" notation to display this visually. In general, we denote particles on I as nodes on a path, with the two fixed particles shaded black. When the inequalities of equation (7.1) become equalities (when particles touch), draw tubes around the respective particles. For example, ●─○─○─○─○─○─○─● corresponds to the configuration

$$0 \leq x_1 \leq x_2 = x_3 = x_4 \leq x_5 \leq x_6 = 1.$$

What about the configuration space $C_2(I)$ of two particles on a line? Since each particle has one degree of freedom (moving along the interval), the space must be 2D. If both particles have no restriction placed on them, the configuration space would be the square $I \times I$, where each

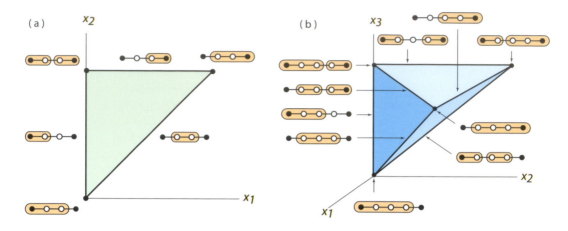

Figure 7.24. Labeling of vertices and edges of $C_2(I)$ and $C_3(I)$.

point (x_1, x_2) in the square keeps track of the position of the two particles on I. However, we know from equation (7.1) that x_1 can never be greater than x_2, resulting in only half of the square, the triangle shown in Figure 7.24(a).

Six chambers of this space (3 edges and 3 vertices) are labeled with the tube notation accordingly. The vertices are points in the configuration space corresponding to places where both particles have collided with fixed particles, leaving no degrees of freedom. The edges, on the other hand, still have one degree of freedom. For the left edge ⬤-○-○-⬤, particle x_2 is free to move, and for the top edge ⬤-○-◉-⬤, particle x_1 is free to move. The third edge ⬤-◉-○-⬤, where both particles have collided with each other, represents positions on I at which this collision could occur. Indeed, all three edges of the triangle can be regarded as copies of the configuration space $C_1(I)$ of Figure 7.23 appearing within $C_2(I)$.

Similarly, $C_3(I)$ is a three-dimensional space, and one can show it to be the tetrahedron pictured in Figure 7.24(b). (In fact, this tetrahedron is one of the six that appear in Figure 1.6(c) in the tetrahedralization of the unit cube.) Once again, the vertices and edges are labeled in this diagram. For higher values of n, the configuration space $C_n(I)$ is a generalization of the tetrahedron called the *n-simplex*. We have encountered this object in Section 1.2 and Section 6.1. We summarize with the following:

Theorem 7.30. *The configuration space $C_n(I)$ of n particles on an interval is the n-simplex.*

Exercise 7.31. *Label the four triangular faces of $C_3(I)$ using the tube notation.*

Exercise 7.32. *Prove that $C_3(I)$ is a tetrahedron.*

Exercise 7.33. *Label the five vertices of the 4-simplex $C_4(I)$.*

Exercise 7.34. *For arbitrary n, how many edges does an n-simplex have? How can we interpret these edges in terms of particle collisions?*

In our study of configuration spaces so far, we have been motivated by motion planning, where our objects have been modeled by moving robots. We now switch our viewpoints slightly, leading to an alternate way of understanding these spaces. From a theoretical physics perspective, we care not just about particle motions but the space of *simultaneous particle collisions*. But what does this mean? In order to explain this abstract concept, we start with a concrete example.

Consider the top-right-most vertex •—o—o—• of Figure 7.24(a) corresponding to the configuration $x_1 = x_2 = 1$. The key question to ask is, *in what ways could this collision have occurred?* Figure 7.25(a) shows a highly zoomed-in view of this vertex in $C_2(I)$. There are several paths drawn here, where the different paths in $C_2(I)$ taken to approach this vertex corresponds to different classes of particle collisions: The approach along the first path shows the three particles colliding, where the latter two particles (x_2 and the fixed particle) have already collided. The third path shows all three particles colliding at the same time, whereas the fifth path shows $x_1 = x_2$, which together collide with the fixed particle. The configuration space $C_2(I)$ displays this collision as a vertex •—o—o—•, giving us a zero-dimensional "point's worth" of information. But as we see from the discussion, there is more information to extract here; indeed, there is a one-dimensional "interval's worth" of possible ways these three particles could have collided. The method by which we reveal this structure is by *truncating* this vertex.[5] Figure 7.25(b) shows the result

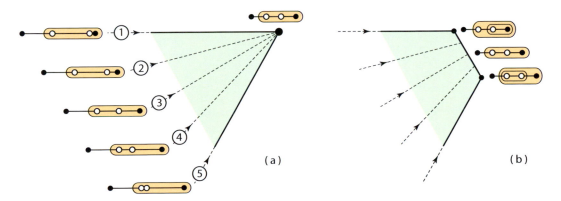

Figure 7.25. A piece of $C_2(I)$, showing (a) paths toward collision and (b) a truncation extracting more information.

---

[5] The actual process is a geometric operation called a *blow-up*.

after truncation. The vertex has now been replaced with an interval, and similarly, the tube labeling of the chambers is now replaced with *nested* tubes.

Figure 7.26(a) shows the triangle $C_2(I)$ with two vertices (marked in red) that contain more information within. Part (b) shows the truncation of these vertices, now replaced by two intervals. This has converted our configuration space of particle motions into a space of particle motions *and collisions*, transforming the triangle into a pentagon. Technically, this pentagon is called the *Fulton-MacPherson compactification* of the configuration space of two particles on an interval. Note that we are not manipulating the environment in which the particles move and interact — this still remains an interval—we are manipulating the configuration space itself!

In a similar manner, we can perform this compactification on the configuration space of three particles on a interval. Here, however, there are several cells whose information needs to be unpacked. Figure 7.27(a) shows two vertices and three edges (marked in red) where three or more collisions have occurred. Although there are other places of collisions, such as the vertex marked by ⬤–o⬤–o–⬤, our interests are in cells with collisions represented by *only one tube*. To obtain our compactified configuration space, we first truncate the two vertices, and then truncate the three edges of the tetrahedron. The resulting space is given by the polyhedron in Figure 7.27(b). The faces of this polyhedron are labeled accordingly.

Notice what used to be a vertex in $C_3(I)$ has now become a pentagon, and what used to be an edge has now become a quadrilateral. This polytope can be seen to have six pentagons and three quadrilaterals for its faces. Indeed, we have seen this before in Figure 3.13(b) as the 3D associahedron! Moreover, the pentagon of Figure 7.26(b) can be viewed

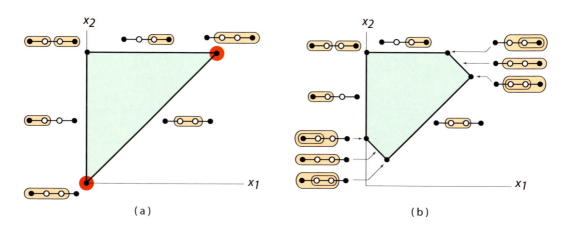

(a) (b)

Figure 7.26. (a) Marked cells of $C_2(I)$ and (b) the resulting polygon after their truncation.

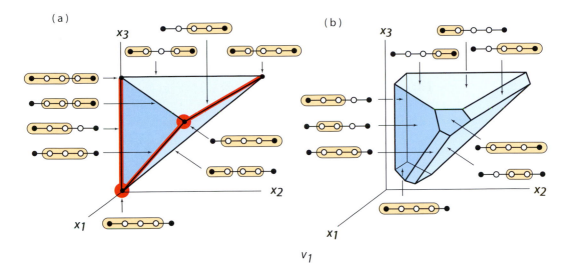

Figure 7.27. (a) Marked cells of $C_3(I)$ and (b) the resulting polyhedron after their truncation.

as the 2D associahedron of Figure 3.12(b). For the general space $C_n(I)$, we have the following beautiful result:

Theorem 7.35. *The Fulton-MacPherson compactification of the config-uration space of n particles on an interval $C_n(I)$ is the n-dimensional associahedron. In particular, the associahedron can be constructed by truncating certain cells of the simplex in increasing order of dimension.*

Figure 7.28 shows a 3D projection of the 4D version of this theorem. Part (a) shows the Schlegel diagram of the 4-simplex. Two vertices

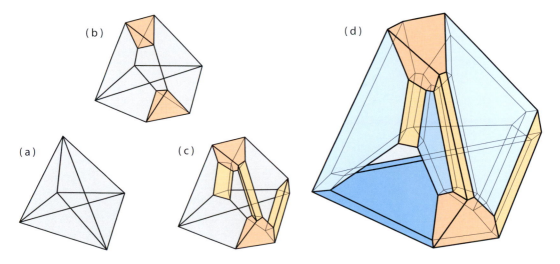

Figure 7.28. (a) The 4-simplex and the truncation of (b) two vertices, (c) three edges, and (d) four faces, resulting in the 4D associahedron.

are then truncated in part (b), similar to the two marked vertices of Figures 7.26(a) and 7.27(a). Three edges are truncated in part (c), and finally four 2D faces are truncated in (d). The resulting object is the 4D associahedron, whose 1-skeleton (the flip graph of a heptagon) is the metal sculpture in Figure 3.14.

Exercise 7.36. *Label the fourteen 3D faces of the 4D associahedron using the tubing notation.*

But how is it that the associahedron appearing in the context of flip graphs should arise in the world of particle collisions? The best way to understand this connection is through Figure 7.29. Here we see a triangulated octagon and its dual tree structure. Notice that the tree is rooted at one of the edges of the polygon. This tree has been redrawn on the right side of the figure, where we see a one-to-one correspondence with nested tubings of 7 particles on an interval. The vertices of the associahedron can thus be seen as triangulations of convex polygons (from the flip graph perspective), or from nested tubings on intervals (from the particle collision perspective). So the triangulated octagon in Figure 7.29 corresponds to the nested parentheses $((1 \cdot 2) \cdot 3) \cdot ((4 \cdot 5) \cdot (6 \cdot 7))$. This is indeed the reason for calling this polytope the *associahedron*, for it captures all the ways of associating particles.

Exercise 7.37. *Prove that triangulations of a polygon with n vertices are in one-to-one correspondence with nested tubings of $n - 1$ particles on the interval. This implies that the Catalan number counts nested tubings as well.*

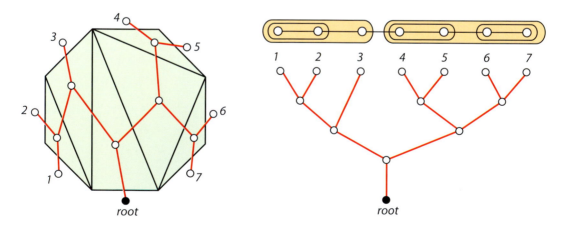

Figure 7.29. A duality between triangulations of polygons, rooted trees, and nested tubings on the interval.

The appearance of the associahedron in the realm of particle collisions on intervals is but a glimpse of one of the worlds in which this wonderful polytope appears. There are numerous generalizations and manifestations of this polytope throughout mathematics, and we close with a glimpse of one. The real moduli space $\overline{\mathcal{M}}_{0,n+3}(\mathbb{R})$ of punctured spheres, which we do not attempt to define here, is an n-dimensional manifold that appears in works ranging from phylogenetic trees in computational biology to string theory in theoretical physics. This moduli space is tiled by $(n + 2)!/2$ copies of the n-dimensional associahedron. Figure 7.30 shows the 3D moduli space $\overline{\mathcal{M}}_{0,6}(\mathbb{R})$ tessellated by 60 copies of the 3D associahedron of Figure 7.27(b). This space can be constructed from the 3D torus (which itself can be obtained by identifying opposites faces of a cube) along with additional truncations and gluings.

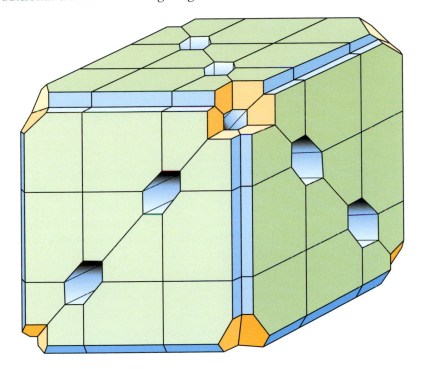

Figure 7.30. The real moduli space $\overline{\mathcal{M}}_{0,6}(\mathbb{R})$ tiled by 60 associahedra.

SUGGESTED READINGS

Erik Demaine and Joseph O'Rourke. *Geometric Folding Algorithms: Linkages, Origami, Polyhedra*. Cambridge University Press, 2007.
Chapter 6 of this book is the source for the locked chains material, including the mentioned graphics morphing result and many open problems.

Joseph O'Rourke. *Computational Geometry in* C. Cambridge University Press, 2nd edition, 1998.
Chapter 8 of this text covers motion planning and robot arm reachability

Colin Adams. *The Knot Book*. American Mathematical Society, 2004.
A wonderfully written and easily accessible book on mathematical knots and their applications, covering the relationship of knots to the topology of surfaces and 3-manifolds.

Don Shimamoto and Catherine Vanderwaart. Spaces of polygons in the plane and Morse theory. *American Mathematical Monthly*, Volume 112, pages 289–310, 2005.
This beautiful paper is our source for polygon configuration spaces. For a classification of singular configurations, see Robyn Curtis and Marcel Steiner, "Configuration spaces of planar polygons" (*American Mathematical Monthly*, Volume 114, pages 183–201, 2007).

Steven LaValle. *Planning Algorithms*. Cambridge University Press, 2006.
This is the 800-page book on motion planning mentioned in Section 7.1. It has an extensive description of probabilistic roadmaps and other sampling-based planning algorithms. For a definitive survey of motion planning complexity results, see Micha Sharir, "Algorithmic motion planning" (in J. E. Goodman and J. O'Rourke, editors, *Handbook of Discrete and Computational Geometry*, chapter 47, pages 1037–1064. CRC Press LLC, 2nd edition, 2004).

Satyan Devadoss. Combinatorial equivalence of moduli spaces. *Notices of the American Mathematical Society*, Volume 51, pages 620–628, 2004.
An introductory, research-level article on particle collisions in the language of algebraic geometry. It shows a larger context in which associahedra appear, such as hyperplane arrangements and the moduli space $\overline{\mathcal{M}}_{0,n}(\mathbb{R})$.

Big-Oh Notation. The *Big-Oh* notation is used to express upper bounds on the growth rate of arbitrary functions, and in particular, upper bounds on the running time of an algorithm. In general, the variable n is used to represent the size of the input to the algorithm, for example, the number of points. The notation ignores constant multipliers and concentrates on the dominate term for large n. Thus an algorithm has running time (or *time complexity*) of $O(n^k)$ if the time as a function of n is less than cn^k for some constant $c > 0$ and for sufficiently large n. For example, suppose that the running time $T(n)$ of an algorithm on a particular processor in particular time units (say seconds) is precisely

$$T(n) = 30.2n \log_2 n + 0.004n^2 + 167.$$

Then this algorithm is $O(n^2)$. The constants are irrelevant as they change from processor to processor. The base of the logarithm is also irrelevant because change of base is absorbed into the constant. And even though the constant on the n^2 term is much smaller than the constant on the $n \log n$ term, for sufficiently large n, the n^2 term will dominate.

Exercise. *For what value of n will* $0.004\,n^2$ *exceed* $30.2\,n \log_2 n$?

Any algorithm that is $O(n^k)$ for some constant k ("constant" means independent of n) is said to have *polynomial* time complexity because it is upper-bounded by some polynomial in n. In some sense, all polynomial-time algorithms are easy because even though the growth rate of, say, n^{100} is fast, it is much slower than, say, 10^n. The latter growth rate is called *exponential*. The functions 2^n, $2^{O(n^2)}$, $n!$ and n^n are all exponential. The table below lists some of the most commonly occurring big-Oh upper bounds and problems with algorithms that achieve those bounds.

Because running time is usually the primary concern, "computational complexity" often means "time complexity." There are circumstances where *space complexity* (the growth rate of the memory required by an algorithm) is quite important, but we do not address this issue in this book.

Big-Omega Notation. The *Big-Omega* notation is the lower-bound equivalent of the big-Oh upper-bound notation. It is especially useful to

Order	Name	Example
$O(1)$	Constant	Adding or multiplying two numbers (of constant size)
$O(\log n)$	Logarithmic	Finding an item in a sorted list by binary search
$O(n)$	Linear	Finding an item in an unordered list
$O(n \log n)$	"$n \log n$"	Sorting a list
$O(n^2)$	Quadratic	Incremental convex hull algorithm
$O(n^k)$	Polynomial	Motion planning for a $(k-1)$-DOF robot arm
$O(c^n)$	Exponential	SET PARTITION via the brute-force algorithm

capture examples that are difficult for an algorithm. An algorithm has running time $\Omega(n^k)$ if the time as a function of n is greater than cn^k for some constant $c > 0$ and for sufficiently large n. Typically one calculates that an algorithm has an upper bound of, say, $O(n^2)$, and then searches for a class of examples that forces the algorithm to consume $\Omega(n^2)$ time. When this is achieved, the running time of the algorithm is called *tight*. Often a gap remains between the upper and lower bounds we can prove, indicating that complete understanding of the algorithm has not been achieved.

Obtaining a lower bound on a problem requires establishing a lower bound for all conceivable algorithms within a certain model of computation. For example, the *decision-tree* model counts comparisons, each of which represents a branch in the tree of decisions, treated as an $O(1)$ computation. Obtaining such a lower bound is considerably more difficult. When a particular algorithm's upper bound matches the problem lower bound, we say the algorithm is *asymptotically optimal*: asymptotically because the relationships only hold for sufficiently large n, and optimal because the matching bounds imply the upper bound cannot be lowered.

NP-Completeness. The *NP-complete* and *NP-hard* problems are a class of problems for which there is no known polynomial-time algorithm. Here P is the class of problems that can be solved in polynomial time; these are the "tractable" ones. And NP is an acronym for "nondeterministically polynomial," which can be understood as those problems whose solutions can be *verified* quickly (in polynomial time). For example, if you are presented with a purported solution to the RULER FOLDING problem of Section 7.3 — a flat configuration whose length is claimed to be at most L — then it is a simple matter to sum the lengths with the appropriate sign (as in Figure 7.14) to verify that this sum is less than or equal to L. However, verifying a given solution quickly is rather different from *finding* a solution in the first place.

Although none of the NP-complete problems are known to require exponential time, no one has found a polynomial-time algorithm for any of them. Moreover, all NP-complete problems are equivalent to one another in the sense that if a polynomial algorithm is found for one, then all can be solved in polynomial time. (NP-hard problems are at least as difficult as NP-complete problems, and may be worse.) As there are more than a thousand known NP-complete problems, none with a polynomial algorithm, the preponderance of evidence is that all are beyond the polynomial-time class P. However, to date this has not been proved. This is the famous $P = NP$ problem, among the six unsolved Millennium Prize Problems.[1] In the absence of a resolution of this question, establishing that a problem is NP-complete or NP-hard is taken as strong evidence that it is intractable.

A few basic problems were proven NP-complete in the 1970s by the pioneers Stephen Cook, Leonid Levin, and Richard Karp. Since then, the general method of proving NP-completeness or NP-hardness is by reducing a problem to another problem that is known to be NP-complete. If the reduction can be accomplished in polynomial time, and the problem is in NP, then that suffices to prove the original problem is also NP-complete. (If the problem cannot be shown to be in NP, the reduction establishes it as NP-hard.) The SET PARTITION problem, discussed in Section 7.3, is one example of a known NP-complete problem.

Another is GRAPH COLORING. We used 3-coloring of triangulations to prove the Art Gallery Theorem in Section 1.3, and it is known that every planar graph can be 4-colored, the famous Four-Color Theorem. But given an arbitrary n-node graph G, the problem of determining its *chromatic number* — the fewest colors needed to color its nodes so that no two adjacent nodes have the same color — is an NP-complete problem. As with most NP-complete problems, there is an obvious exponential algorithm: try all possible 2-colorings, then all possible 3-colorings, and so forth. Because n colors suffice, this tedious brute-force enumeration will eventually find the chromatic number. But for this problem, and for all NP-complete problems, no one has either found a polynomial-time algorithm, nor proved that exponential time is truly needed.

UNSOLVED PROBLEM 29 $\hspace{2cm} P = NP$

Either prove that $P \neq NP$ by showing that some NP-complete problem needs exponential time, or prove that $P = NP$ by finding a polynomial-time algorithm for some NP-complete problem.

---

[1] http://www.claymath.org/millennium/. The seventh prize problem was the Poincaré conjecture, now settled as described in Section 5.6.

SUGGESTED READINGS

Thomas Cormen, Charles Leiserson, Ron Rivest, and Cliff Stein. *Introduction to Algorithms*. MIT Press, 3rd edition, 2009.

Now in its third edition, this massive (1300-page) textbook is both readable and comprehensive.

Robert Sedgewick. *Algorithms in* C++. Addison-Wesley, 1992.

A popular textbook that emphasizes implementations. Available in several language editions: C, C++, Java.

Jon Kleinberg and Éva Tardos. *Algorithm Design*. Pearson Education, 2006.

A text that emphasizes design as much as analysis, and connects to real-world applications at every turn.

Figure 3.25. From "Minimum-weight triangulation is NP-hard," *Journal of the ACM (JACM)*, **55**:2 (2008) 1–29. Used with permission of the authors, Wolfgang Multzer and Günter Rote.

Figures 3.26 and 3.27. From J. Danciger, S. Devadoss, D. Sheehy, "Compatible triangulations and point partitions by series-triangular graphs," *Computational Geometry: Theory and Applications*, **34** (2006) 195–202. Used with permission from Elsevier.

Figure 5.6. From J. Erickson and D. Eppstein, "Raising roofs, crashing cycles, and playing pool: Applications of a data structure for finding pairwise interactions," *Discrete and Computational Geometry*, **22** (1999) 569–92. Used with permission from Springer Science+Business Media.

Figures 5.18. From "2D Minkowskisums," Chapter 22 of *The CGAL Manual*. `http://www.cgal.org/Manual/`. Used with permission of the author, Ron Wein.

Figures 6.24, 6.27, 6.29. From E. Demaine and J. O'Rourke, *Geometric Folding Algorithms: Linkages, Origami, Polyhedra*. Cambridge Univ. Press, 2007. (Figures 23.6, 22.4, 22.17, 24.1). Used with permission from Cambridge University Press.

Figure 7.8. Used by permission from OwiRobots, Omnico Group. `http://www.owirobots.com/`.

Figure 7.18. From "Using motion planning to study protein folding pathways," *Journal of Computational Biology* **9**:2(2002) 149–68. Used with permission from Mary Ann Liebert, Inc.

Figures 7.24, 7.26, 7.27, 7.28, 7.30. From S. Devadoss, "Combinatorial equivalence of real moduli spaces," *Notices of the American Mathematical Society* **51** (2004) 620–28. Used with permission from the American Mathematical Society.